CONSUMER DURABLE CHOICE
AND THE DEMAND FOR ELECTRICITY

CONTRIBUTIONS
TO
ECONOMIC ANALYSIS

155

Honorary Editor:
J. TINBERGEN

Editors:
D. W. JORGENSON
J. WAELBROECK

NORTH-HOLLAND
AMSTERDAM · NEW YORK · OXFORD

CONSUMER DURABLE CHOICE AND THE DEMAND FOR ELECTRICITY

JEFFREY A. DUBIN

California Institute of Technology
Pasadena, CA 91125
U.S.A.

1985

NORTH-HOLLAND
AMSTERDAM · NEW YORK · OXFORD

ISBN: 0 444 87766 5

Publishers:
ELSEVIER SCIENCE PUBLISHERS B.V.
P.O. Box 1991
1000 BZ Amsterdam
The Netherlands

Sole distributors for the U.S.A. and Canada:
ELSEVIER SCIENCE PUBLISHING COMPANY, INC.
52 Vanderbilt Avenue
New York, N.Y. 10017
U.S.A.

Library of Congress Cataloging in Publication Data

Dubin, Jeffrey A.
    Consumer durable choice and the demand for electricity.

    (Contributions to economic analysis ; 155)
    Includes index.
    1. Electric power consumption--Mathematical models.
2. Household appliances, Electric--Energy consumption--
Mathematical models. I. Title. II. Series.
HD9685.A2D83 1985   333.79'3212'0724   85-7010
ISBN 0-444-87766-5

Printed in The Netherlands

Introduction to the series

This series consists of a number of hitherto unpublished studies, which are introduced by the editors in the belief that they represent fresh contributions to economic science.

The term "economic analysis" as used in the title of the series has been adopted because it covers both the activities of the theoretical economist and the research worker.

Although the analytical methods used by the various contributors are not the same, they are nevertheless conditioned by the common origin of their studies, namely theoretical problems encountered in practical research. Since for this reason, business cycle research and national accounting, research work on behalf of economic policy, and problems of planning are the main sources of the subjects dealt with, they necessarily determine the manner of approach adopted by the authors. Their methods tend to be "practical" in the sense of not being too far remote from application to actual economic conditions. In addition they are quantitative rather than qualitative.

It is the hope of the editors that the publication of these studies will help to stimulate the exchange of scientific information and to reinforce international cooperation in the field of economics.

The Editors

Contents

Introduction

In the years from 1947 to 1972, the United States experienced an almost sevenfold increase in the use of electricity. The early 1970s brought the intertwined problems of dwindling oil resources, increased dependence on oil imports, and a heightened need for a consensus on national energy policy. However, increasing concern over the safety of nuclear power mitigated the trend toward pervasive electrification and the nation's all-electric future.

The need to quantify the responsiveness of electricity utilization to various energy policies rose rapidly in the energy-turbulent 1970s. This need was recognized all the way down to the level of homeowners, who became concerned with efficiency and costs of alternative heating and cooling systems. Of course homeowners who had witnessed an increase in their energy budget from 26 percent in 1972 to 33 percent in 1980 knew all too well that the composition of their appliance stock greatly influenced their usage.[1]

Energy researchers also noted the importance of durable stocks in the energy demand process.[2] Yet, only in very recent attempts have econometric simulation models allowed policy scenarios simultaneously to affect appliance holdings and resultant usage. In one direction are aggregate studies that fit average appliance saturations to the time trend of income, prices, and other socioeconomic variables. This approach is best exemplified in the modeling efforts of Hirst and Carney (1978). Other aggregate-based studies are extensively reviewed in Hartman (1978, 1979b).[3]

In contrast to the aggregate studies, several attempts to model jointly the demand for appliances and the demand for fuels by appliance have

[1]See "Annual Report to Congress, Volume Two: Data," U.S. Department of Energy, Energy Information Administration Report DOE/EIA-0173- (80)/2 (April 1981, pp. 9).

[2]Classical studies of aggregate electricity consumption given appliance stocks are Houthakker (1951), Houthakker and Taylor (1970), and Fisher and Kaysen (1962). A number of other studies postulate an adaptive adjustment of consumption to long-run equilibrium, which can be attributed to long-run adjustments in holdings of appliances; see Taylor (1975).

[3]The Hartman review describes both single-fuel and interfuel substitution models. Among the single-fuel demand studies based on aggregate data, Hartman includes Acton, Mitchell, and Mowill (1976 and 1978), Anderson (1973), Chern and Lin (1976), Hartman and Werth (1981), Mount, Chapman, and Tyrell (1973), Wilder and Willenborg (1975), and Wilson (1971).

been completed using cross-sectional microlevel survey data.[4] The use of disaggregated data is desirable as it avoids the confounding effects of either misspecification due to aggregation bias or misspecification due to approximations in rate data.

Either approach has a common objective in modeling household energy consumption patterns from which to evaluate conservation and load management policies. For example, can we evaluate the welfare and distributional impacts of proposed government policies to decontrol the price of natural gas? How rapidly do consumers respond to rising energy prices? What are the differences between the energy consumption of owner-occupiers and tenants? What are the implications for public information programs that provide energy efficiency labeling and building and appliance standards? Does the marketplace offer sufficient incentives for consumers to pursue appropriate levels of conservation? What actions should government take, if any, to encourage conservation? Can we quantify the long-run and short-run responses to policy actions and describe the time path of conservation?

To begin to understand these issues, we present in Figure 1 a conceptualization of residential energy consumption process. Household demographics, household income, fuel prices, equipment prices, and climate are inputs to a residential choice process that determines appliance and dwelling characteristics. Included in appliance characteristics are fuel types, capacities, efficiencies, and holdings. Included in dwelling characteristics are structure type, size, and thermal integrity.

The appliance investment decision is characterized by several unique features. First, many durable goods in the home are installed at the time of house construction. In this category are the heating, ventilating, and air-conditioning systems, as well as the water-heating unit. While individuals have the opportunity of voluntary replacement or retrofitting, reliability and ease of monitoring combined with quite large transactions costs lead to substantial inertia in these durable goods. Cross-sectional surveys indicate relatively infrequent changes in the household physical plant during the preceding three to five years with observed movement caused principally by appliance failure.

---

[4]Cross-sectional studies with this structure are McFadden, Kirschner, and Puig (1978), the residential forecasting model of the California Energy Conservation and Development Commission (1979), the microsimulation model developed by Cambridge Systematics/West for the Electric Power Research Institute described in Cambridge Systematics/West (1979), Goett (1979), and Goett, McFadden, and Earl (1980).

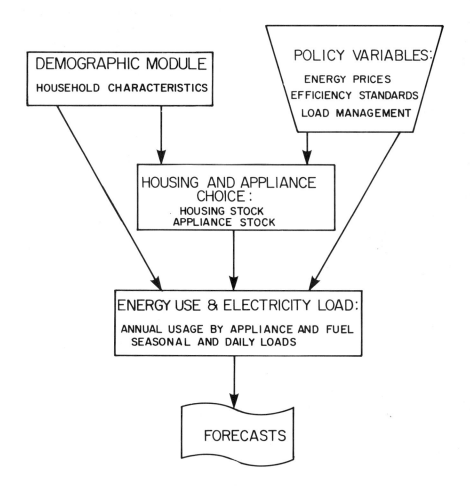

Figure 1

Our second observation is that the appliance investment decision is much more flexible in new construction. National construction estimators find that the cost of newly installed air conditioning in existing dwellings may be as much as three times greater than that in new construction. Therefore, it seems sensible to concentrate on the durable selection problem in new construction. Moreover, conservation policy often focuses on new construction given the potential for energy savings.

Our analysis of the appliance investment decision assumes that households respond to the cost and performance characteristics of alternative durable equipment types that depend upon the attributes of the dwelling and upon the environment in which the durable operates. Initial or capital cost is incurred at the time of dwelling construction and determined by capacity, but may be strongly influenced by the availability of certain fuel types and the joint cost characteristics of installed equipment. Operating cost or the cost-per-unit service is the cost required to maintain the dwelling at a specific internal temperature. In order to "size" the dwelling, i.e., determine required capacities and to calculate expected energy loads, we employ an engineering thermal load model.

Engineering thermal load models calculate the amount of heat entering and leaving the residence for each hour of the day and are capable of determining loads for space-conditioning end-uses. These calculations require detailed input including data on the physical, thermal, and operational characteristics of the dwelling, as well as location-specific hourly temperature data. These models are highly specialized to determine both static and dynamic heat transfer.

The engineering thermal load technique has been found quite accurate when detailed information on building characteristics exists. The methodology incorporates complex nonlinear relationships between weather, building characteristics, and thermal loads and thus provides significant *a priori* information in statistical analysis. Furthermore, the technique may be used to assess the impact of conservation and load management programs that affect building characteristics, as well as to provide estimates of system load at extreme weather conditions.

The thermal load technique is combined with billing cycle data in our study in two unique ways. First, we use the thermal model to estimate billing cycle load on a household-by-household basis. In this approach, two households with equivalent building characteristics facing identical weather patterns are predicted to have the same energy demand. In reality, we realize that the demands may vary significantly between otherwise identical households due to differences in income, household size, activity patterns, and the cost of energy. Departures from the engineering estimates are due to socioeconomic sensitivity in the rate of appliance stock utilization. Secondly, we use the engineering thermal load techniques to estimate the cost of comfort. Here the estimated change in energy input required to effect a one-degree change in ambient temperature is multiplied by the marginal price of the fuel input.

Having determined the appliance and housing stock, households react to policy and market variables such as energy prices, tax credits, rebate offers, efficiency standards, etc., to determine energy usage by appliance and by fuel type. Economic analysis suggests that the demand for consumer durables arises from the flow of services provided by durable ownership. Energy, for example, is consumed as an input to the household production process to provide heating, cooling, and other services. The utility associated with a consumer durable is then best characterized as indirect. Durables may vary in capacity, efficiency, and versatility, and of course will vary correspondingly in price. Although durables differ, the consumer will ultimately utilize the durable at an intensity level that provides the "necessary" service. In the case of the household, usage is determined by the installed heating, ventilating, and air-conditioning (HVAC) system and by interior and exterior temperatures.

The optimization problem is best characterized as a dynamic stochastic program in which consumers maximize the expected utility of durable utilization given realized energy prices and expectations regarding the distribution of operating costs induced by random exogenous weather and unanticipated price changes. In the spirit of the theory, consumers must weigh the alternatives of each appliance whose technology is embodied in the characteristics of dwelling against expectations of future use, future energy prices, and current financing decisions.

Given that consumers have a somewhat limited ability to process all available information, it might be expected that a fully rational optimizing model for behavior would have only little advantage in predictive power over simpler characterizations. The specification of practical econometric demand systems for fuel usage must by necessity compromise some aspects of the optimization problem posed above. These models presuppose that consumers can detect prevailing marginal fuel rates in the presence of automatic appliances, billing cycle variations, and limited information on appliance operating characteristics. More fundamentally, there is the assumption that the shares of appliance portfolios in recent construction provide information on consumer preferences independently of portfolio decisions made by contractors and sellers.

This assumes that the structure of supply and demand encourages sellers to act as de facto agents for buyers so that the distinction between construction to stock and construction to order disappears. Provided that the energy characteristics of dwellings can be audited costlessly, economic rents should accrue to builder/contractors who provide the mix of energy systems most preferred by buyers. For example, an unanticipated change in the relative prices of electricity and natural gas should shift consumer demand toward dwellings with HVAC systems that use the cheaper fuel. In the short run, suppliers of such systems experience positive economic rents, while in the longer run, market forces tend to equalize returns with new construction providing evidence of the equilibrated market shares of distinct appliance portfolios.

For the purposes of forecasting, the residential energy consumption process is assumed to be recursive. In the first stage, a housing decision is made. Conditional on the housing decision, appliance portfolios are chosen by the household, and finally, energy demand is determined conditional on the choice of appliance stock. This recursive structure may be used in policy analysis. For example, consider a change in the building code that requires all new dwellings to meet a baseline thermal integrity standard through wall and ceiling insulation. Increased thermal integrity in the housing stock alters the structure of operating and capital costs of available heating and cooling systems. Changes in expected operating and capital costs produce predictable shifts in the saturations of alternative heating and cooling systems. Furthermore, the demand for fuels by appliance then reflect the increased thermal integrity of the dwelling and the resultant change in the marginal cost of heating and cooling services.

For the purposes of estimation, however, it must be recognized that the consumer's demand for durables and their use are related decisions. Econometric specifications that ignore this fact lead to biased and inconsistent estimates of price and income elasticities. Dubin and McFadden (1984) test the exogeniety of appliance dummy variables typically included in electricity demand equations. Their approach derives an indirect utility function that is consistent with the specification of a partial demand equation. The indirect utility function is used to predict portfolio choice while the demand equation predicts conditional electricity usage.[5] Employing a logistic discrete-choice model of all electric and all natural gas space and water heating systems combined with conditional demand for electricity, Dubin and McFadden reject the hypothesis

---

[5]Related work in the area of discrete/continuous econometric systems is given in McFadden (1979a), Duncan (1980a), Duncan (1980b), Duncan and Leigh (1980), Hay (1979), King (1980), Lee and Trost (1978), McFadden and Winston (1981), and Hausman and Trimble (1981).

that unobserved factors influencing portfolio choice are independent of the unobserved factors influencing intensity of use. The econometric model of consumer durable choice and utilization employed by Dubin and McFadden consists of simultaneous equations with dummy endogenous variables (Heckman, 1978, 1979) and may be thought of as a switching regression with a structure analyzed by Lee (1981), Goldfeld and Quandt (1972, 1973, and 1976), Maddala and Nelson (1974 and 1975), and Fair and Jaffee (1972).

Our approach recognizes that neither purely technological nor purely behavorial models are likely to provide accurate forecasts of energy demand. Instead, we explicitly model the physical constraints imposed by the durable technology and emphasize the role of consumer behavorial response in energy consumption that operates through the choice of appliances and appliance characteristics. Here we analyze the residential demand for electricity and natural gas conditioned on the choice of space heating, water heating, and central and room air-conditioning choice utilizing the National Interim Energy Consumption Survey (NIECS) 1978 survey of 4081 households. The model developed is intended to have the flexibility to be included in a large microsimulation forecasting system (such as the Residential End-Use Energy Policy System (REEPS)).[6] We further extend the theoretical development of durable choice and utilization and examine the hypothesis of simultaneity between appliance choice and electricity and natural gas demand.

In Chapter One, we develop the theory of durable choice and utilization. The basic assumption is that the demand for energy is a derived demand arising through the production of household services. The technology that provides the household service is the appliance durable. Durable choice is then associated with the choice of a particular technology from a set of alternative technologies. Using results from household production theory, we derive econometric systems that capture both the discrete-choice nature of appliance selection and the determination of continuous conditional demand.

Chapter Two develops a thermal load model specifically designed for application to household survey data. While there are many models available to calculate heating and cooling requirements, most are designed to be used by contractors and architects on individual dwellings

---

[6]For details concerning the implementation of a large-scale energy forecasting model, such as REEPS, the reader should consult Goett (1979) and Cambridge Systematics/West (1979).

where detailed measurements are available.[7] This thermal model makes reasonable assumptions about dwelling characteristics and operating practices that are not coded in typical survey data while utilizing all information about insulation levels, window counts, etc., that is readily available. The approach also simplifies the task of providing detailed weather data and is able to process summary measures such as temperature means and extremes. The methodology is superior to the use of simple degree-day measures while allowing calculations in large samples of dwellings.

Chapter Three describes the estimation of a discrete-choice model for room air conditioning, central air conditioning, space heating, and water heating. The form of the appliance choice model results from the assumption that the unobserved components of utility have a generalized extreme value distribution. A particular form of this distribution is considered that implies that the choice of room air conditioning given the choice of central air conditioning is independent of the choice of the space heating system given the choice of central air conditioning. Water heating fuel choice is assumed to depend only on the choice of space heating system.

Chapter Four reviews the theory of price specification and considers the comparative static analysis of demand subject to a declining block rate schedule. We further investigate the statistical endogeneity of prices whose construction requires utilization of the observed consumption level, and determine price specification within a sample of 744 households surveyed in 1975 by the Washington Center for Metropolitan Studies (WCMS). We finally consider the construction of marginal prices using the WCMS data and monthly billing data from NIECS.

Chapter Five considers an efficiency comparison of various two-stage consistent estimation techniques applied to a multiequation switching regime model with known but possibly endogenous regimes. This class of models covers the demand system estimated in Chapter Six as well as the system of Dubin and McFadden (1984) and Heckman (1979). Asymptotic distributions are derived for each estimator using the methods of Amemiya (1978a, 1979).

---

[7]Examples are NBSLD, developed by the National Bureau of Standards; DOE-2, developed by Lawrence Berkeley Laboratory for the Department of Energy; BLAST, developed by the Army Civil Engineering Research Laboratory; and the residential building model developed by the Ohio State University for the Electric Power Research Institute.

Chapter Six presents the estimation of the demand for electricity and natural gas. Consistent estimation procedures are used in the presence of possible correlation between the dummy variables indicating appliance holdings and the equation error. Estimation is based on monthly billing data matched to each household in the NIECS survey. The monthly billing data provides an excellent time profile of usage that permits the determination of seasonal effects.

The main text is followed by two technical appendices. The first appendix describes the processing of the NIECS data and the creation of an appended NIECS data base. It further describes the creation of marginal electricity and natural gas prices based upon the theory of Chapter Four.

The second appendix presents the calculation of various conditional moments in the generalized extreme value family. These results extend the analysis given in Dubin and McFadden (1984) for the case of discrete continuous econometric systems where discrete choice is assumed logistic. Finally, this appendix provides the conditional expectations used in selectivity corrections of dummy endogenous variable systems in which the probabilistic choice system is nested logistic.[8]

---

[8]The nested logit model is described in McFadden (1978, 1979b, and 1981b).

Acknowledgments

This book summarizes my research efforts in energy economics during the years 1981 through 1984. In 1981, a detailed study of durable choice and utilization was begun at the Massachusetts Institute of Technology. This work culminated in my dissertation under the direction of Daniel McFadden. This work owes its intellectual origins to Dan and has largely benefited from our collaborative efforts. In the development of the thesis, I must further acknowledge the comments of Ernst Berndt and Franklin Fisher, who served on my doctoral committee. Financial support during this period was provided by National Science Foundation grants to McFadden. The resources of the M.I.T Energy Laboratory are also gratefully acknowledged.

The years 1983 and 1984 involved the substantial revision of the original dissertation and the completion of new empirical work. Of the original presentation, Chapters 1 and 4 and Appendix A have changed the least. Chapter 2, which discusses an energy thermal model for single-family residences, was developed initially by McFadden and me and circulated under McFadden and Dubin (1982), "A Thermal Model for Single-Family Owner-Occupied Detached Dwellings in the National Interim Energy Consumption Survey." At the California Institute of Technology, the thermal model was modified to predict cooling loads more precisely and to be compatible with survey data on households in the Pacific Northwest (PNW) prepared by the Bonneville Power Administration.

Chapter 3 received the greatest attention during 1983. At this time I began an analysis of durable choice for residences in the PNW and a comparison of these results with those obtained using the National Interim Energy Consumption Survey. Steven Henson provided continued help in the preparation of the PNW data and has more generally stimulated many discussions on the role of durable choice in energy demand modeling. Our discussions and later collaboration under Bonneville Power Administration grant DE-AI79-83BP13579 on energy utilization in the PNW have helped my own presentation of the results in Chapter 6. The more technical sections, Chapter 5 and Appendix B, have benefited from the comments of an anonymous referee to North-Holland who made several suggestions that simplified the presentation.

My experience in the preparation of this manuscript has been often frustrating but sometimes quite enjoyable. Dale Jorgenson, as editor to

North-Holland, made the original suggestion to use the UNIX photo-typesetting system TROFF available at Caltech for photocomposition. Starting with very little knowledge of typesetting, I have now gained first-hand experience in preparing a difficult and long text.

During 1983 and 1984, during which the majority of the typesetting was accomplished, several individuals provided their assistance. Barbara Calli typed the manuscript into the Caltech word-processing system. Lisa Doermann served ably as copy editor and proofreader. Edith Huang, of the Caltech computer center, provided invaluable and patient help with the technical problems involved in using TROFF. The Caltech Environmental Quality Laboratory provided the services of a graphic artist to aid in the preparation of the figures. Computing resources and general research funds were provided by an EXXON corporation grant to the Enviromental Quality Laboratory.

On a personal note, I want to thank my wife, Jackie, for her encouragement and support. Only she knows how much time and sacrifice were involved in the completion of this book. The book is dedicated to my parents, to whom I owe the greatest debt and appreciation. While I've attempted to connect the various players to the many pieces that compose this study, all errors are indelibly my own.

CHAPTER 1

Consumer durable choice and utilization[1]

1.1. Introduction

In this chapter we consider models of consumer durable choice, hold-ings, and utilization. Examples are drawn primarily from the literature on electricity demand and appliance choice, but much of the exposition is consistent with a wider realm of household behavior. For instance, the methodology could be used to develop a model of household automobile choice and utilization without substantive modification.

Consumer durable models are usefully classified by their treatment of durable utilization in addition to the frequent distinction between hold-ings and purchase. Broadly speaking, a purchase model analyzes the decision to acquire a durable stock while a holdings model attempts to explain how the stock evolves during its economic life.[2]

In general, any model of consumer durable choice should consider:

1) the distinction between the decision to purchase a stock of durable goods and the decision to hold or replace that stock,

[1] I gratefully acknowledge the useful comments of Thomas Cowing, Peter Navarro, and Clifford Winston.
[2] Examples of pure holdings models are Diewert (1974), who uses the classical stock-flow model to analyze the demand for money over time, and Griliches (1960), who uses a stock-adjustment model to estimate the demand for farm equipment. Pure purchase or choice models are considered by Chow (1957), Cragg and Uhler (1970), and Cragg (1971) in the context of the demand for automobiles, Li (1977) for housing choice, and Joskow and Mishkin (1977) for utility fuel choice. Appliance purchase models are considered by McFadden, Kirschner, and Puig (1978).

Examples of holdings and utilization models are the classical stock-utilization studies of aggregate electricity consumption given appliance stocks by Houthakker (1951b), Houthakker and Taylor (1970), and Fisher and Kaysen (1962). Stock-adjustment models with utilization are treated in the work of Balestra and Nerlove (1966) on the demand for natural gas.

Purchase or choice models for durable goods that jointly consider utilization are very re-cent. Dubin and McFadden (1984), Hartman (1979a), and Hausman (1979) all consider discrete-choice models of appliance ownership and corresponding utilization.

1

2)   the inherent "discreteness" of durable goods, e.g., while additional cooling may be provided by an individual-room air conditioner, available units offer only fixed ranges of capacity,

3)   the imperfection or nonexistence of rental markets for durable resale,

4)   the sizable transaction and installation costs often connected with the decision to retrofit or upgrade a durable stock,

5)   the intertemporal utility maximization problem that results from the inherent dynamic choice of a durable stock and the utilization of that stock over its lifetime,

6)   the characterization of any solution to be based on information available to the consumer at the time the decision is made, the modifications to that solution as new information becomes available, e.g., technological innovation or change in the relative costs of alternative fuels, and

7)   the link between a durable good and the technology that it often embodies.

Unfortunately, previous literature has failed to incorporate all of these crucial points in a consistent model of durable choice behavior. For example, the classical holdings model of consumer durables as presented in Diewert (1974) assumes perfect foresight, perfect rental markets, and a flow of services resulting from a stock of durable goods that depreciates but may be augmented continuously. This capital-theoretic framework fails to integrate the purchase decision with the decision to utilize or change the durable stock. The initial choice of durable stock with given features is crucially important, however, because the realization of levels or rates of change of key economic variables that differ from the consumer's *ex ante* predicted values may make the *ex ante* optimal durable choice *ex post* undesirable. Faced with low resale values of his durable stock and inaccessible resale markets, the consumer would not be expected to change his durable stock often and perhaps only when very large changes in utility had occurred. Furthermore, prices of durable goods reflect their capitalized rents and hence tend to have values that become significant fractions of consumers' budgets. The resolution of financing large initial setup costs may directly affect durable choice when some consumers' access to capital markets is limited. This may indirectly affect the choice of other economic goods and thus affect consumer welfare.

The importance of initial purchase is derived from the notion that once a durable stock is purchased, it will remain intact for many years. The classical model de-emphasizes the purchase decision by allowing "putty-putty" flexibility in durable stocks.

It would be unfair to say that the classical model cannot treat aspects of transaction costs and limited rental markets. Such factors may be incorporated into stock-flow models but invariably surface in their effects on the "user cost of capital." A change in the user cost of capital induces an immediate and continuous response in the desired level of durable stock.

As an alternative to the classical model, this chapter considers the general discrete-choice model. The discrete-choice model assumes that the purchase, holding, and replacement decisions correspond to differences in utility values crossing threshold levels. The decision to change the level of durable holdings is viewed as a discrete movement from one durable portfolio combination to another. This change is typically costly and occurs infrequently for the usual consumer.

The discrete and classical models of individual choice behavior differ in that the former does not assume that the stock of durable goods can be changed continuously. Thus differences between desired and actual stocks are not instantaneously or adaptively actualized as in the classical model. Finally, depreciation itself is often a stochastic phenomenon that represents durable failure and necessitates a repair or replacement decision on a highly intermittent basis. These distinctions are potentially important because they may imply rather different choice behavior by consumers. A comparison of the predictive abilities of the discrete-choice approach with the classical model of durable choice awaits our empirical results.

The bulk of this chapter then is concerned with rigorously developing a theoretical and econometric framework for analyzing durable choice from a discrete-choice perspective. We begin (in Section 1.2) with the development of the discrete-choice approach by considering two examples.

The first example explores the economic and engineering characterization of residential heating and comfort. Here we define the concept of an "energy price of comfort" and indicate the method by which it may be constructed empirically. The second example motivates the characterization of durable selection as the choice of a particular technology for producing household services that yield direct satisfaction to the consumer. This link to household production theory relaxes the assumed proportionality relationship between flows and stocks in the classical model.

In Section 1.3, we seek conditions affecting technology and preferences under which household production of durable services follows a two-stage plan. In the first stage, consumers determine optimal production service levels; in the second stage, they choose input combinations that produce these services at minimum cost. Section 1.4 introduces econometric models of discrete choice and utilization with explicit attention given to the link with the theoretical model and the treatment of stochastic components.

In particular we examine a demand system with additively-separable disturbances that is consistently derived from utility-maximization theory. The joint model of appliance choice and utilization is the basis for the empirical analysis that follows in later chapters.

1.2. Consumer durable choice and appliance technology

The demand for energy by the household is a derived demand arising from the production of household services. The technology that provides household services is embodied in the household appliance durable. Thus to understand the residential demand for energy, we must understand the residential demand for durable equipment.

A neoclassical consumer will base decisions regarding appliance purchase, replacement, and retirement on the life-cycle capital and operating costs of alternative appliance portfolios. With uncertainty about future fuel prices, appliance operating characteristics, and imperfect intertemporal and used-appliance markets, each consumer will evaluate life-cycle costs on the basis of his expectations and discount rate. Further, in the absence of complete rationality, the consumer is likely to respond to simpler market indicators than fully articulated life-cycle cost.

The first econometric problem in analyzing appliance choice is that the components of life-cycle appliance cost are usually not all observable. Cross-sectional surveys often provide insufficient information to determine vintage, purchase price, or efficiency of individual appliances. To the extent that these unmeasured components are uniform over the sample, or vary independently of observed operating costs, it is possible to estimate response to costs consistently. This case is plausible for a contemporaneous cross section but not for retrospective data on historical behavior. A second difficulty is that contemporary energy prices may be a poor indicator of the operating-cost expectations of a household. This is a particularly acute problem for historical appliance decisions. An adequate analysis of the interaction of price expectations and appliance purchases requires collection of the historical energy price data known to consumers at the decision point. A third, and more fundamental, difficulty in analyzing appliance portfolio decisions lies in the question of the interaction of supply and demand.

Except for retrofit decisions, the choice of many consumer durables is fixed at the date of construction. Because most dwellings are constructed to stock rather than to order, it would appear that appliance portfolio decisions are largely made by contractors, and that purchase prices of houses adjust to clear existing stocks of houses with various appliance portfolios. Thus, consumer preferences would be detectable primarily in relative profit margins, or economic rents, on different housing types rather than in appliance portfolio shares. Contractors, however, will shift the mix of new housing construction toward types with the highest profit margin. The flow of new housing into the market will tend to restore equality between profit rates for different housing types by adjusting supply to meet the shares demanded at current life-cycle appliance portfolio costs in new construction. If new housing supply is sufficiently elastic, the overall housing market will be brought into equilibrium, and the analysis of appliance portfolio shares as a function of life-cycle costs in new construction will provide information on consumer preferences independently of contractors' tastes.

Examination of purchase shares alone, however, may lead to overestimates of the sensitivity of appliance shares to costs, because cost shifts inducing modest long-run shifts in relative stocks will typically cause extreme short-run shifts in investment shares. If housing supply is not perfectly elastic, an increase in relative operating cost for a housing type will induce little change in portfolio share and an unobserved decrease in house price for this housing type, and may lead to an underestimate of the sensitivity of consumer choice to cost.

With these issues in mind, we consider in some detail the household selection of a space-heating system. This decision may arise as a result of the installation of a heating system in new construction, as part of a technological upgrading of the existing stock (the "retrofit" decision), or from replacement due to existing system failure. Observation suggests that households choose a temperature profile during a 24-hour period, which they attempt to attain using their heating system. For some households this may involve setting the thermostat at one temperature during the day and at another level at night. Other households rely on thermostat timers, others simply the "feel" of the coldness in the air.

Specifically, we let $u[t, Z]$ denote the utility derived from consumption of a vector of goods Z in an environment with ambient temperature t. It is reasonable to assume that utility is increasing in t up to a temperature t^*, which provides blissful comfort. Below t^* occupants feel too cool and above t^* feel too hot. If heating were a free good, consumers would set their thermostats at t^*. But because heating to an interior temperature t^* requires a costly energy input, there is a trade-off between the comfort of the ambient space and the price of obtaining this comfort.

Following Brownstone (1980) and Hausman (1979), we assume that the utility function $u[t, Z]$ is separable in comfort and goods consumption. Furthermore, we suppose that $u[t]$, the utility derived from ambient temperature t, takes the linear form $u[t] = -\alpha[t^*-t]$ for $\alpha > 0$ and $t \leqslant t^*$. Let $F[t]$ denote the cumulative distribution for the number of days in which the daily mean temperature is less than or equal to t. Utility during the heating season from thermostat setting τ is:

$$u[\tau] = \int_{-\infty}^{\tau} -\alpha(t^*-\tau)\, F'(t)\, dt + \int_{\tau}^{t^*} -\alpha(t^*-t)\, F'(t)\, dt \qquad (1)$$

The first integral assumes that comfort is constant at the level $(t^*-\tau)$ degrees per hour when outside temperature is below the thermostat level τ. The second integral assumes that comfort increases proportionally to increases in temperature below the bliss temperature point. It is simply demonstrated that equation (1) has an interpretation measured in degree-days of heating. From equation (1):

$$u[\tau] = -\alpha \left[(t^*-\tau)F(\tau) + t^*[F(t^*) - F(\tau)] - \int_{\tau}^{t^*} tF'(t)\, dt \right]$$

$$= -\alpha \left[t^*F(t^*) - \tau F(\tau) - \int_{\tau}^{t^*} tF'(t)\, dt \right]$$

$$= -\alpha \left[\left(t^*F(t^*) - \int_{-\infty}^{t^*} tF'(t)\, dt \right) - \left(\tau F(\tau) - \int_{-\infty}^{\tau} tF'(t)\, dt \right) \right]$$

$$= \alpha[H(\tau) - H(t^*)]$$

where $H(t)$ denotes total heating degree-days measured at base t, i.e.,

$$H[t] = \int_{-\infty}^{t} (t-s)F'(s) \, ds = tF(t) - \int_{-\infty}^{t} sF'(s)ds$$

The degree to which a given housing structure loses heat to the colder outside is related directly to the size of the various exposed surfaces and their conductivity to heat flow as well as the absolute temperature differential. Insulation in the walls and ceiling and the presence of storm windows all lower the overall thermal conductivity of the housing shell and hence the demands on a heating system to maintain a given comfort level.

Suppose that the Btuh heating required to maintain interior temperature τ with exterior temperature t is given by the function $Q(\tau-t)$. Then the seasonal heating load resulting from thermostat setting τ is:

$$B[\tau] = \int_{-\infty}^{\tau} \max[\ Q(\tau-t), \ 0\ ] F'(t) \, dt \tag{2}$$

The problem of maximizing the utility function $U[\tau, Z]$ subject to a budget constraint allocates income I between expenditure on goods Z and on fuel $(p_i/e_i)B(\tau)$ where p_i is the price of fuel i and e_i is the efficiency of the heating system using fuel i. We write:

$$\underset{\tau, \, Z}{\text{maximize}} \ \ U[\tau, Z] \ \text{subject to} \ (p_i/e_i)B[\tau] + Z \leqslant I \tag{3}$$

for which the Lagrangian (with multiplier ξ) is:

$$L = U[\tau, Z] + \xi[I - Z - (p_i/e_i)B(\tau)] \tag{4}$$

The first order conditions are:

$$L_\tau = U_\tau - \xi(p_i/e_i)B'(\tau) = 0 \quad \text{and} \tag{5}$$

$$L_Z = U_Z - \xi = 0 \qquad \text{so that:} \tag{6}$$

$$U_\tau / U_Z = (p_i/e_i)B'(\tau) \tag{7}$$

We see from (7) that the price of comfort depends on the level of comfort. It is possible to reformulate the optimization problem using an appropriately defined rate structure premium. Let τ^* denote the solution to (7) so that $(p_i/e_i)[B(\tau^*)-B'(\tau^*)\tau^*]$ is the rate structure premium adjustment. The equivalent standardized problem is then:

$$\text{maximize}_{\tau, Z} \ U[\tau, Z] \ \text{subject to} \ (p_i/e_i)B'(\tau^*)\tau$$
$$+ Z \leqslant I - (p_i/e_i)[B(\tau^*) - B'(\tau^*)\tau^*] \tag{8}$$

The indirect utility associated with equation (8) is a function of I and the price of comfort $(p_i/e_i)B'(\tau^*)$. The thermal model discussed in Chapter 2 is used to estimate the price of comfort for alternative HVAC systems. The procedure approximates the derivative $B'(\tau^*)$ by calculating the change in seasonal utilization associated with a one-degree change in the thermostat setting. In our empirical work we ignore the rate structure premium adjustment to I of equation (8).

Consider the added complexity of heating-system capacity. As the temperature differential between inside and outside increases, the limit of an installed system for providing delivered Btu's of heat may be reached. Recommended construction practice suggests that a space-heating system should provide adequate heating capacity against all but the coldest 1 percent of the heating season.[3] But it is not purely an engineering decision that determines capacity. Required capacity is in fact a function of the levels of insulation, of air-infiltration, and of the behavioral patterns of the household.

Given a selected heating-system capacity, households then choose among available technologies and delivery systems. For example, space heating is commonly provided by central forced air, wall units, hot water radiators, etc. Each system is available at a corresponding capital cost. In choosing a given space-heating system, consumers face an economic

[3]Chapter 2 provides a detailed account of the construction of a thermal-load model for single-family residences.

decision in which they compare the initial nonutility of purchasing the capital equipment with the future utility of the heating services provided by its operation.

If the class of household durable systems were continuous, then it would be appropriate to apply the methodologies of Rosen (1974) and Wills (1977) and model appliance selection as household production using differentiated capital goods.

For our purposes it will suffice to extend the household production paradigm to permit both discrete and continuous choices. We illustrate the approach through a second example.

Assume a one-period world in which consumers have the choice of two technologies for providing an identical end-use service. The isolated choice of a gas or an electric clothes dryer for providing a given service level, e.g., pounds of dry clothes per day, fits into this category.

Suppose that the alternative technologies are given by $Y_1^j = f_j(x_j; a_j)$ with purchase prices $H_j(a_j)$, $j = 1,2$.

Vectors x_1 and x_2 represent inputs to the respective technologies and may be purchased at prices p_1 and p_2. The parameters a_1 and a_2 are vectors of attributes that define the production technologies. These attributes might, for example, measure limits on drying capacity or efficiency. Conditioning production on the parameters a_j in the function f_j corresponds to the notion of a restricted technology. Note that the continuously variable parameters determine purchase price through the hedonic function H_j, but recognize that technological varieties are finite.

We assume that preferences are representable by a single-period utility function $U[Y_1, Y_2]$ where Y_1 is the end-use service level provided by either of the alternative technologies and Y_2 is a transferable numeraire or Hicksian commodity.

The consumer's decision problem is to make an *ex ante* technology choice recognizing that *ex post*, income I will be allocated among expenditures on input commodities and all other goods to achieve maximal utility in goods and services.

The indirect utility corresponding to the choice of technology j is:

$$V[I - H_j[a_j], P_j] = \max_{x_j,\, a_j} U[Y_1^j, Y_2] \quad \text{subject to:}$$

$$Y_1^j = f_j[x_j;\, a_j] \quad \text{and} \quad p_j x_j + H_j[a_j] + Y_2 \leqslant I \tag{9}$$

Consumers choose technology 1 if and only if:

$$V[I - H_1, p_1] \geqslant V[1 - H_2, p_2] \tag{10}$$

This implies that unconditional indirect utility is given by:

$$V^*[I - H_1, I - H_2, p_1, p_2] = \max_j V[I - H_j, p_j] \tag{11}$$

In this example, the *ex ante* choice between technologies is discrete. Either technology 1 is purchased or technology 2 is purchased. This choice has an immediate income response through the purchase price H_j.

The budget set in final goods and services corresponding to the first technology is:

$$c_1 = \{ (Y_1, Y_2) \in \mathbf{R}_+^2 \mid Y_1 = f_1(x_1; a_1) ;$$

$$p_1 x_1 + Y_2 + H_1(a_1) \leqslant I; x_1 \geqslant 0 \} \tag{12}$$

When the production function $f_1(x_1; a_1)$ is invertible, (12) may be written:

$$c_1 = \left\{ (Y_1, Y_2) \in \mathbf{R}_+^2 \mid p_1 f_1^{-1}[Y_1; a_1] + Y_2 \leqslant I - H_1 \right\} \tag{13}$$

where $f_1^{-1}[Y_1; a_1]$ denotes the assumed non-negative quantity of input x_1 necessary to produce service level Y_1 given parameters a_1.

Assume that the technology is smooth so that the marginal rate of substitution and its rate of change can be calculated on the boundary of c_1. From (13):

$$dY_2/dY_1 = -p_1/f_1'(x_1; a_1) < 0 \qquad \text{and} \tag{14}$$

$$d/dx_1[dY_2/dY_1] = \frac{f_1''(x_1; a_1)\, p_1}{[f_1'(x_1; a_1)]^2} \quad < 0 \tag{15}$$

where we have assumed that f is strictly increasing and concave in its first argument. The set c_1 is illustrated in Figure 1.

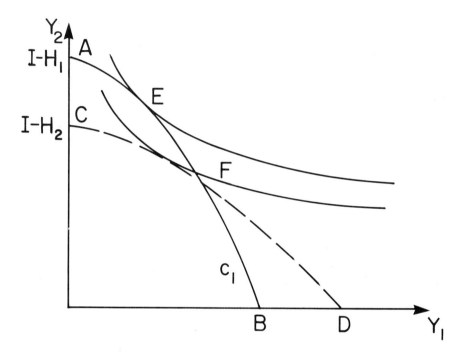

Figure 1.1

Strict convexity of the budget set is implied by (15). The budget set corresponding to the second technology is the area beneath the dotted line connecting points C and D. Figure 1 illustrates a situation in which maximal utility in final goods and services is achieved at points E and F corresponding to *ex ante* choice of technologies 1 and 2 respectively. In this example, maximal utility would be achieved through choice of technology 1.

The Lagrangian for (9) (with multipliers λ_1 and λ_2) is:

$$L = U[Y_1^1, Y_2] + \lambda_1[Y_1^1 - f_1(x_1; a)]$$

$$+ \lambda_2[I - p_1 x_1 - H_1 - Y_2] \tag{16}$$

The first-order conditions yield the tangency solution:

$$\frac{-U_1[Y_1^1, Y_2]}{U_2[Y_1^1, Y_2]} = \frac{\lambda_1}{\lambda_2} = \frac{-p_1}{f_1'(x_1; a_1)} \tag{17}$$

Equation (17) equates the marginal rate of substitution between end-use services, Y_1^1, and all other goods, Y_2, to the marginal cost of producing Y_1^1.

Application of the envelope theorem to equation (16) shows that $L_I = \lambda_2$ and $L_{p_1} = -\lambda_2 x_1$ so that:

$$\frac{-V_p[I - H_1, p_1]}{V_I[I - H_1, p_1]} = \frac{-L_2[I - H_1, p_1]}{L_1[I - H_1, p_1]} = x_1 \tag{18}$$

Here we see that Roy's identity holds for input or intermediate commodities.[4]

Thus far, we have assumed strict concavity of the production function $Y_1^1 = f(x_1; a_1)$, which implies the strict convexity of budget constraint set c_1. When the production function is in fact linear in x_1, a fixed-coefficient technology results. In this case the boundary of c_1 is flat, and we may define a service price for end-use consumption that is constant. Furthermore, linearity in the input good x_1 ensures that the *average efficiency of production* defined by the service level achieved per quantity of input utilized is constant.

The appropriate extension of the concept of average efficiency to cases in which production exhibits decreasing returns to scale is the notion of marginal efficiency. We define the *marginal efficiency of production*

---

[4]Dubin and McFadden (1984) use this result along with simple assumptions about technology to derive a consistent econometric choice and utilization system.

resulting from input x as the marginal product of x given parameters a. Thus the electrical efficiency of providing cooling-degree hours of air conditioning will depend on climate, usage levels, insulation-unit capacity, etc. The quantity $p_1/f_1'(x_1; a_1)$ in (17) is the end-use service price for $Y_1^!$. We see that the end-use service price or marginal cost of $Y_1^!$ is the price of input commodity x_1 divided by the marginal efficiency of x_1.

These examples have illustrated how the consumers' durable choice problem can be represented in terms of the optimal choice of technology subject to financial and technological constraints. In the next section we derive conditions under which the separability in utility implied by appliance-production technologies permits a consistent two-stage or "tree" budget program. Under the two-stage budgeting procedure, consumers first determine optimal production service levels and then choose input combinations that produce these service levels at minimum cost.

1.3. Appliance technology and two-stage budgeting

The examples presented in section 1.2 make clear that household energy demand is a derived demand for basic fuel inputs to appliance technologies.[5] Here we discuss the *ex post* utilization of a given appliance stock and examine conditions under which the optimal allocation of inputs may be separated by appliance type. Section 1.4 considers *ex ante* utilization as it affects indirect utility and therefore the selection of alternative appliance portfolios.

Suppose that utility is given as a function of n end-use service goods and a vector of non-produced commodities:

$$U(x) = U[f_1(x_1; a_1), f_2(x_2; a_2), ..., f_n(x_n; a_n), x_{n+1}] \tag{19}$$

where:

$f_j(x_j; a_j)$ = production of end-use service Y_j

x_j = vector of input commodities for production of end-use service j

---

[5]The reader is referred to Becker (1965) and Muth (1966) for other characterizations of the household as a production unit.

a_j = parameters that affect production

x_{n+1} = vector of non-produced goods.

Equation (19) does not impose any restrictions upon the set of commodities that yield direct utility. It is apparent, however, that the production functions $f_j(x_j; a_j)$ *generically* separate the commodities x_j. The partition is termed generic because the same physical commodity is often an input for several distinct technologies. This interpretation regards electricity used as an input to clothes drying as distinct from electricity used as an input for space heating, yet both inputs are priced identically. Total electricity demanded is the sum of electricity demanded in each end-use. We suppose that the input commodities x_j are available at prices p_j and that p_{n+1} is the price vector for all other goods x_{n+1}. The budget constraint for traded commodities is:

$$\sum_{j+1}^{n} p_j x_j + p_{n+1} x_{n+1} \leqslant I \tag{20}$$

where I denotes the level of predetermined total expenditure (less the cost of the chosen appliance portfolio).

Conditional on the choice of technologies, consumers must allocate resources to maximize (19) subject to (20). Let $c_j(Y_j, p_j; a_j)$ be the cost function dual to the production function $f_j(x_j; a_j)$. We can recast the optimization problem using the cost functions as:

$$\max U[Y_1, Y_2, ..., Y_n, x_{n+1}]$$

$$\text{subject to } \sum_{j-1}^{n} c_j(Y_j, p_j; a_j) + p_{n+1} x_{n+1} \leqslant I \tag{21}$$

By direct analogy to Gorman's proposition (Gorman (1959)), we see that necessary and sufficient conditions for a consistent two-stage budgeting solution to (21) in which consumers first determine optimal service levels and then choose input combinations that produce these service levels at minimum cost require that production be homothetic.[6] A

---

[6]This implication of the Gorman proposition is discussed in Blackorby, Lady, Nissen, and Russel (1970).

stronger condition, employed by Muellbauer (1974) and Pollak and Wachter (1975), assumes that the production technologies exhibit constant returns to scale. For the purposes of this discussion, we adopt this assumption but note that the essential features of the argument are unchanged provided a new utility indicator is defined that is consistent with renormalized production functions.[7] Under constant returns to scale in production, the cost functions have the simple form $c_j(Y_j, p_j; a_j) = \theta_j[p_j; a_j] \cdot Y_j$, where the unit-cost functions $\theta[\cdot ; a_j]$ are homogeneous of degree one. The assumption of constant returns to scale in production does not appear to limit the generality of our approach. It may, however, be necessary to limit the range of production through the parameters a_j.

The optimization problem in (21) becomes:

$$\max U[Y_1, Y_2, ..., Y_n, x_{n+1}]$$

$$\text{subject to } \sum_{j=1}^{n} \theta_j [p_j; a_j] \cdot Y_j + p_{n+1} \cdot x_{n+1} \leqslant I \tag{22}$$

from which indirect utility is:

$$V[\theta_1(p_1; a_1), \theta(p_2; a_2), ..., \theta_n (p_n; a_n), p_{n+1}, I] \tag{23}$$

where V is dual to U in (19).

We see from (23) that indirect utility satisfies a price partition that corresponds to the commodity partition assumed in (19). The crucial element of the derivation is that the utility function U in (19) is homothetically separable in appliance technologies.

The functions $\theta_j(p_j; a_j)$ have a straightforward interpretation as the unit costs of producing end-use service j.

From our discussion in Section 1.2, we recognize the unit cost of space-heating services, for example, as the product of the efficiency-adjusted price of electricity and the marginal change in utilization resulting from a one-degree change in the thermostat setting. Thus

---

[7]A production function $f(x)$ is homothetic when $f(x) = g(h(x))$, with g monotonic and h linearly homogeneous. If the utility function is given by $u[f(x)]$, then $\bar{u}(z) = (u \cdot g) (z)$ is consistent with the linearly homogeneous function $z = h(x)$.

$\theta_{sh}(p_e; a_{sh}) = (p_e/eff_{sh}) \cdot B'(\tau^*; a_{sh})$ where eff_{sh} is the efficiency of the space-heating system and $B'(\tau^*; a_{sh})$ is the marginal change in energy utilization. The parameters a_{sh} include all structural characteristics of the residence that affect heating-system performance.

Given the specification of unit-cost functions, Shephard's Lemma determines optimal input factors:

$$x_j = [\partial\theta_j(p_j; a_j)/\partial p_j] \cdot Y_j \tag{24}$$

Equation (24) demonstrates that the *input-to-service ratios* x_{jk}/Y_j for input k are independent of service level. Let V_j and V_I denote the derivatives of (23) by the j-th service price and by income respectively. Roy's identity applied to (23) determines optimal service levels in the first stage of the two-stage budget procedure:

$$Y_j = \frac{-V_j[\theta(p_1; a_1), ..., \theta(p_n; a_n), p_{n+1}, I]}{V_I[\theta(p_1; a_1), ..., \theta(p_n; a_n), p_{n+1}, I]} \tag{25}$$

To derive the total demand for a given input, we use (24) and (25) to determine input utilization by end-use and then sum across end-uses. By way of example, suppose that each technology uses electricity and that the price of electricity appears as an argument in the functions $\theta(p_j; a_j)$. Total demand for electricity, x_e, satisfies:

$$x_e = -\sum_{j-1}^{n} \frac{\partial\theta_j(p_j; a_j)}{\partial p_e} \cdot \left[\frac{V_j}{V_I}\right] \tag{26}$$

Econometric estimation of the electricity demand equation (26) could proceed once a selection of the indirect utility V is made. Alternatively, it is possible to begin with an econometric specification for Y_j in (25) and then derive a consistent indirect utility function. We pursue the econometric specification of (26) in the next section.

1.4. Econometric specification for models of durable utilization

We presented in Section 1.3 a two-level utilization procedure in which service levels, Y_j, are determined by equation (25) and optimal input combinations required to produce Y_j are determined by (24). Econometric specification for this system requires explicit functional forms for indirect utility, V, and for service levels Y_j. As Roy's identity connects V with Y_j through (25), it is often possible to specify a parametric form for demand and then solve a partial differential equation to find a compatible indirect utility function. This methodology has been successfully applied by Hausman (1979, 1981b), Burtless and Hausman (1978), and Dubin and McFadden (1984) for individual demand equations. We now consider the recovery of an indirect utility function from a *system* of demand equations as required by (25). We follow Dubin and McFadden (1984) and assume that demand is linear in real income I and additive with a function of real prices:

$$Y_j = \beta_j I + m_j(p_1, p_2, ..., p_n) \qquad j = 1,2, ..., n \qquad (27)$$

By Roy's identity we may write the first equation in this system as:

$$\frac{-\partial V/\partial p_1}{\partial V/\partial I} = \beta_1 I + m_1(p_1, p_2, ..., p_n) \qquad (28)$$

We apply the implict function theorem and write (28) in differential form as:

$$-[\beta_1 I + m_1(p_1, ..., p_n)] \, dp_1 + dI = 0 \qquad (29)$$

Application of the integrating factor $\mu(p_1, p_2, ..., p_n, I) = e^{-\beta_1 p_1} \cdot g(p_2, ..., p_n)$ transforms (29) into an exact differential equation with solution:[8]

---

[8]The solution of exact differential equations and the use of integrating factors are discussed in any elementary text on ordinary differential equations. See Boyce and DiPrima (1969, pp. 39-43) for discussion.

$$V(p_1, p_2, ..., p_n, I) = e^{-\beta_1 p_1} g(p_2, ..., p_n) [I + M(p_1, ..., p_n)] \qquad (30)$$

where:

$$M(p_1, p_2, ..., p_n) = \int_{p_1} e^{\beta_1(p_1 - t)} m_1(t, p_2, ..., p_n) \, dt \qquad (31)$$

Note that (31) satisfies:

$$\partial M / \partial p_1 - \beta_1 M = -m_1 \qquad (32)$$

Roy's identity applied to (30) for the second commodity implies:

$$Y_2 = \frac{-\partial V / \partial p_2}{\partial V / \partial I}$$

$$= \frac{-e^{-\beta_1 p_1} g(p_2, ..., p_n) M_{p_2} - e^{-\beta_1 p_1} [I + M] g_{p_2}}{e^{-\beta_1 p_1} g(p_2, ..., p_n)} \qquad (33)$$

$$= -M_{p_2} - [I + M] g_{p_2}/g \quad \text{where} \quad M_{p_2} = \frac{\partial M}{\partial p_2}$$

Comparing (33) with (27) we must have $-g_{p_2}/g = \beta_2$ and $-M_{p_2} + \beta_2 M = m_2(p_1, ..., p_n)$. Proceeding similarly for commodities $j = 3, ..., n$, we find:

$$V(p_1, p_2, ..., p_n, I) = \left[e^{-\Sigma \beta_j p_j} \right] \cdot (I + M(p_1, p_2, ..., p_n)) \qquad (34)$$

where the function M satisfies the restrictions:

$$\beta_j M - M_{p_j} = m_j \quad \text{for } j = 1, 2, ..., n \qquad (35)$$

The restrictions in (35) imply a relationship among the m_j that must be satisfied if (34) is consistent with (27). These restrictions are identical to symmetry of the Slutsky substitution matrix as we now demonstrate. Consider (35) for $j = 1,2$:

$$\beta_1 M - M_{p_1} = m_1 \rightarrow e^{-\beta_1 p_1} (\beta_1 M - M_{p_1}) = e^{-\beta_1 p_1} \cdot m_1 \tag{36}$$

and:

$$\beta_2 M - M_{p_2} = m_2 \rightarrow e^{-\beta_2 p_2} (\beta_2 M - M_{p_2}) = e^{-\beta_2 p_2} \cdot m_2 \tag{37}$$

From (36) and (37) we have:

$$\frac{\partial}{\partial p_1} \left[e^{-\beta_1 p_1} \cdot M \right] = -e^{-\beta_1 p_1} \cdot m_1 \quad \text{and} \tag{38}$$

$$\frac{\partial}{\partial p_2} \left[e^{-\beta_2 p_2} \cdot M \right] = -e^{-\beta_2 p_2} \cdot m_2 \tag{39}$$

from which follow:

$$\frac{\partial}{\partial p_1} \left[e^{-\beta_1 p_1 - \beta_2 p_2} \cdot M \right] = -e^{-\beta_1 p_1 - \beta_2 p_2} \cdot m_1 \quad \text{and} \tag{40}$$

$$\frac{\partial}{\partial p_2} \left[e^{-\beta_1 p_1 - \beta_2 p_2} \cdot M \right] = -e^{-\beta_1 p_1 - \beta_2 p_2} \cdot m_2 \tag{41}$$

Equating the mixed partials of (40) and (41) we have:

$$\frac{\partial}{\partial p_2} \left[e^{-\beta_1 p_1 - \beta_2 p_2} \cdot m_1 \right] = \frac{\partial}{\partial p_1} \left[e^{-\beta_1 p_1 - \beta_2 p_2} \cdot m_2 \right] \quad \text{or} \tag{42}$$

$$\partial m_1 / \partial p_2 - \beta_2 m_1 = \partial m_2 / \partial p_1 - \beta_1 m_2 \tag{43}$$

By Slutsky symmetry we have:

$$\partial Y_1/\partial p_2 + \partial Y_1/\partial I \cdot Y_2 = \partial Y_2/\partial p_1 + \partial Y_2/\partial I \cdot Y_1 \tag{44}$$

which implies:

$$\partial m_1/\partial p_2 + \beta_1 Y_2 = \partial m_2/\partial p_1 + \beta_2 Y_1 \quad \text{or} \tag{45}$$

$$\partial m_1/\partial p_2 + \beta_1(m_2 + \beta_2 I) = \partial m_2/\partial p_1 + \beta_2(m_1 + \beta_1 I) \tag{46}$$

so that:

$$\partial m_1/\partial p_2 + \beta_1 m_2 = \partial m_2/\partial p_1 + \beta_2 m_1 \tag{47}$$

Comparing (43) with (47) we find that conditions (35) are equivalent to symmetry of the substitution matrix. Additional integrability restrictions (homogeneity, summability, non-negativity, and negative quasi-semi-definiteness) are imposed on M by the requirement that $V(p_1, p_2, ..., p_n, I)$ be an indirect utility function.[9] Under these conditions, it is not difficult to verify that equation (34) is in generalized Gorman polar form.[10]

Combining equations (26) and (27), we see that the demand for electricity takes the form:

$$x_e = \sum_{i-1}^{n} \frac{\partial \theta_j(p_j; a_j)}{\partial p_e} \cdot \left[\beta_j I + m_j(\theta_1, ..., \theta_n) \right] \tag{48}$$

A virtue of the Gorman polar form is that it is compatible with a partial demand system with additive disturbances. To demonstrate this property we modify equation (34) so that:

[9]The integrability problem is discussed in Chipman, Hurwicz, Richter, and Sonnenschein (1971).

[10]The generalized Gorman polar form may be written $V[\tilde{p}, I] = (I - a(\tilde{p}))/ b(\tilde{p})$ where $I > a$ and a and b are homogeneous of degree one, concave, and non-decreasing in $\tilde{p}$. Application of Roy's identity yields: $Y_i = a_i(\tilde{p}) + (b_i(\tilde{p})/ b(\tilde{p})) \cdot (I - a(\tilde{p}))$ where $a_i(\tilde{p}) = \partial a(\tilde{p})/ \partial p_i$ and $b_i(\tilde{p}) = \partial b(\tilde{p})/ \partial p_i$.

$$M(p_1, p_2, ..., p_n) = M^*(p_1, p_2, ..., p_n) + \sum_{j=1}^{n} p_j \eta_j + \xi_1 \tag{49}$$

where ξ_1 and η_j, $j = 1,2, ..., n$ are random components. Then equation (35) implies:

$$m_j(p_1, p_2, ..., p_n) = \beta_j M^* + \beta_j \sum_{j=1}^{n} p_j \eta_j + \beta_j \xi_1 - \frac{\partial M^*}{\partial p_j} - \eta_j$$

$$= m_j^* + \beta_j \sum_{j=1}^{n} p_j \eta_j + \beta_j \xi_1 - \eta_j \tag{50}$$

where $\quad m_j^* = \beta_j M^* - \partial M^*/\partial p_j$

Combining equations (48) and (50) we have:[11]

$$x_e = \sum_{j=1}^{n} \psi_j \left[\beta_j I + m_j^*(\tilde{\theta}) + \beta_j \sum_{j=1}^{n} \theta_j \eta_j + \beta_j \xi_1 - \eta_j \right]$$

$$= \sum_{j=1}^{n} \psi_j [\beta_j I + m_j^*(\tilde{\theta})] + \xi_1^* \tag{51}$$

where:

$$\psi_j = \partial \theta_j(p_j; a_j)/\partial p_e$$

$$\xi_1^* = \sum_{j=1}^{n} (\psi^* \theta_j - \psi_j)\eta_j + \psi^* \xi_1 \quad \text{and}$$

$$\psi^* = \sum_{j=1}^{n} \psi_j \beta_j$$

[11]Here we adopt the notation $\tilde{\theta} = (\theta_1, \theta_2, ..., \theta_n)$.

To complete the econometric system we propose a probabilistic choice system over the discrete technologies. Let i^* denote an individual portfolio among J possible technological varieties. Following McFadden (1981b), we write the utility of alternative i^* as $U^{i^*} = V^{i^*} + \varepsilon^{i^*}$ where V^{i^*} is the indirect utility of the alternative and ε^{i^*} denotes unobserved components of utility. The probability that portfolio $i*$ is chosen satisfies:

$$P_{i*} = \text{Prob}\left[U^{i^*} \geqslant U^{j^*}, \forall j* \mid j* \neq i* \right]$$

$$= \text{Prob}\left[V^{i^*} + \varepsilon^{i^*} \geqslant V^{j^*} + \varepsilon^{j^*}, \forall j* \mid j* \neq i* \right]$$

$$= \text{Prob}\left[\varepsilon^{j^*} - \varepsilon^{i^*} \leqslant V^{i^*} - V^{j^*}, \forall j* \mid j* \neq i* \right] \tag{52}$$

Note that the estimation of the econometric system consisting of equations (51) and (52) should account for the endogeneity of variables indicating the portfolio choice $i*$. A detailed comparison of available estimation techniques is considered in Chapter 5.

Chapter 2 is devoted to the construction of a residential energy thermal model that permits empirical determination of operating and capital costs for alternative heating and cooling systems. Chapter 3 then implements the probabilistic choice model in (52) under specific assumptions on the unobserved components ε^{i^*}.

CHAPTER 2

A heating and cooling load model for single-family detached dwellings[1]

2.1. Introduction

The National Interim Energy Consumption Survey (NIECS) and the Pacific Northwest Residential Energy Survey (PNW) are clustered, random samples of households interviewed between 1978 and 1980. These surveys were designed to report household equipment holdings and energy consumption by fuel, as well as selected household and dwelling characteristics. To study the economic determinants of equipment and usage behavior, it is necessary first to describe the economic environment in which behavior is determined. This chapter carries out the construction of heating-ventilating-air conditioning (HVAC) physical characteristics and costs for alternative systems available to single-family, owner-occupied households.

The approach of this chapter is to construct a very simple thermal model of representative dwellings with characteristics corresponding to those available in typical energy survey data. This model is used to estimate heating and cooling capacity requirements, energy usage, and physical characteristics for households in the NIECS and PNW surveys. Cost data from Means (1981) are then used to estimate the capital and operating costs of 19 alternative HVAC configurations for the actual thermal integrity of the building shell and for two alternative thermal standards.

[1]This chapter revises and extends a coauthored draft with Daniel McFadden entitled "A Thermal Model for Single-Family Owner-Occupied Detached Dwellings in the National Interim Energy Consumption Survey," M.I.T. Energy Laboratory Discussion paper No. 25, MIT-EL 82-040WP. Our research was supported in part by NSF Grant No. 80-16043-DAR, Department of Energy, under Contract No. EX-76-A-01-2295, Task Order 67, and the Environmental Quality Laboratory of the California Institute of Technology. We wish to acknowledge a substantial contribution to this research by Thomas Cowing, who provided weather and location data for NIECS households, and to Jean Yates-Rimpo, who provided assistance with the Pacific Northwest data.

2.2. Thermal modeling principles

Heating and cooling system design capacities of a dwelling are determined by the rates of heat transfer between the interior and exterior
under extreme weather conditions. Conduction and infiltration are the
dominant modes of transfer in winter; radiation is important in summer.
Heating capacity calculations normally assume steady-state thermal conditions, while cooling calculations take account of inertial (flywheel)
effects.

The approach to capacity calculation adopted here follows engineering
practice, as detailed in ASHRAE (1977, 1978, 1979), Anderson (1973),
Khashab (1977), and Streeter (1966). Application of these principles to
the NIECS and PNW households requires a number of assumptions and
model simplifications due to incomplete data on dwelling characteristics.

A dwelling may be pictured as a box with walls of varying thermal
resistances to conduction of heat, as depicted in Figure 2.1.

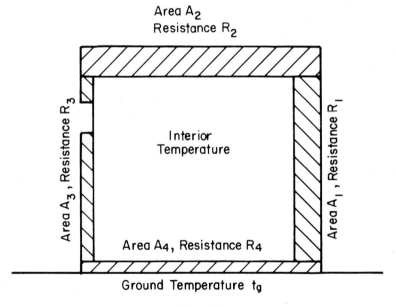

Figure 2.1

The net heat loss by conduction from the dwelling in Btuh is the sum of
the losses through each area, which equals the area times the temperature differential divided by the resistance (measured in ft^2 × °F / Btuh):

$$\frac{\text{Net conduction}}{\text{heat loss}} = \frac{A_1\,(T_i-T_e)}{R_1} + \frac{A_2\,(T_i-T_e)}{R_2} \tag{1}$$

$$+ \frac{A_3\,(T_i-T_e)}{R_3} + \frac{A_4\,(T_i-T_g)}{R_4}$$

This formula omits boundary effects due to the exterior temperature gradient near the ground surface and at the interface of surfaces with different resistances. In practice, these effects are usually small and can be neglected. When a correction is required, it can be calculated using elementary circuit theory. The method is illustrated in Figure 2.2 for the example of a wall in a heated basement that has an exterior temperature gradient. The wall can be represented by a network of nodes connected by conductors with resistances equal to the thermal resistances of the intervening material. The precision of the calculation is improved by increasing the number of nodes. In the example, R_1 and R_2 are resistances of wall material from the dwelling interior to the wall center and from the wall center to the exterior, while R_7 is a resistance to vertical conduction. Heat flow along a link equals the temperature difference between the link nodes times the cross-sectional area represented by the link, divided by the resistance of the link. In thermal equilibrium, net heat flow into an interior node is zero. These conditions plus the interior and exterior temperatures define a system of linear equations in the node temperatures and link heat flows. In the example, these equations are:

$$\begin{bmatrix} R_1^{-1}+\lambda R_7^{-1} + R_2^{-1} & -\lambda R_7^{-1} & 0 \\ -\lambda R_7^{-1} & R_3^{-1}+R_4^{-1}+\lambda R_7^{-1}+\lambda R_8^{-1} & -\lambda R_8^{-1} \\ 0 & -\lambda R_8^{-1} & R_5^{-1}+R_6^{-1}+\lambda R_8^{-1} \end{bmatrix}$$

$$\times \begin{bmatrix} T_i-T_1' \\ T_i-T_2' \\ T_i-T_3' \end{bmatrix} = \begin{bmatrix} R_2^{-1}(T_i-T_1) \\ R_4^{-1}\,(T_i-T_2) \\ R_6^{-1}(T_i-T_3) \end{bmatrix} \tag{2}$$

where the cross-sectional areas associated with R_1 and R_6 are assumed to equal 1 and the cross-sectional areas associated with R_7 and R_8 are assumed to equal λ. Then:

$$H = \frac{1}{3} \text{(wall area)} \left[\frac{T_i - T_1'}{R_1} + \frac{T_i - T_2'}{R_3} + \frac{T_i - T_3'}{R_5} \right] \tag{3}$$

is the heat loss through the wall.

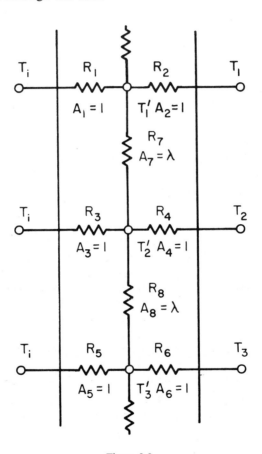

Figure 2.2

ASHRAE (l977, Chap. 22) provides data on the resistances of various construction materials. These permit calculation of resistances of standard construction. The NIECS/PNW data do not indicate type of wall construction, whether the roof is pitched, whether there is a basement, or whether walls and roof are light or dark. For purposes of estimating design capacities, we therefore make the following assumptions:

1. Exterior walls are of standard frame construction with exterior wood siding.

2. The dwelling has a pitched roof with an unheated attic with natural ventilation.

3. There is an *unheated* basement that is primarily below grade.

4. Roof and walls are dark in color.

It should be noted that variations in construction will cause substantial variations in the thermal performance of dwellings. Hence, the model developed here should not be expected to predict precisely dwelling-to-dwelling variations in thermal performance, even for dwellings satisfying the four assumptions above. On the other hand, construction standards tend to compensate for differences in resistance usually arising in dwellings that do not fit these assumptions. For example, construction standards for flat roofs generally call for insulation between roofing and sheathing, which offsets the loss of resistance provided by an attic. Similarly, masonry walls with lower resistance than frame walls are normally more heavily insulated, as are slab floors in comparison with construction over an unheated basement. Consequently, we do not expect resistance calculations based on the assumptions above to be systematically biased for alternative types of construction. Table 2.1 gives the resistance of standard frame exterior wall construction. Table 2.2 gives the resistance of ceiling and roof for flat and pitched roofs. A later analysis incorporating the effects of solar radiation will combine these resistances into a single, effective ceiling-roof resistance. Table 2.3 gives the resistance of glass, excluding radiation effects. Table 2.4 gives floor resistance.

Table 2.1

Exterior wall resistance

Material	R-value
Outside Surface (15 mph windspeed)[1]	0.17
Wood Siding	0.87
Building Paper	0.06
Sheathing (0.5 plywood)	0.62
Air Space (framing)[2]	0.94 (4.38)
Gypsum Wallboard	0.45
Inside Surface	0.68
Resistance of portion of wall with framing (10 for 16 o.c. framing)	7.23
Resistance of portion of wall without framing or insulation	3.79
Resistance of portion of wall with insulation (R-value = I)	2.85 + I
Average resistance of wall without insulation[3]	3.98
Average resistance of wall with insulation (R-value = I)[3]	$\dfrac{12.85 + I}{0.9394 + 0.0138I}$

[1]Surface resistance decreases with wind speed. The ASHRAE design standard is 15 mph winter windspeed and 7.5 mph summer windspeed, the latter giving a surface resistance of 0.25.

[2]Standard 2" × 4" framing is assumed, giving an air space of 3.75 . Without insulation the R-value of the air space is 0.94. At typical insulation R-value of 3.0/inch, the R-value of light insulation (1.5) is 4.5, and of heavy insulation (3.5) is 10.5. The R-value of the wood framing members is 1.17 per inch.

[3]The average resistance of the wall satisfies:

$$R_{average}^{-1} = R_{framing}^{-1} \left(\begin{smallmatrix}proportion\\framing\end{smallmatrix}\right) + R_{other}^{-1} \left(\begin{smallmatrix}proportion\\other\end{smallmatrix}\right).$$

Source: ASHRAE (1977), 22.13-22.22, particularly Tables 4A,G,I,K.

Table 2.2

Ceiling and roof resistance

Flat Roof and Ceiling	Heating R-value	Cooling R-value
Outside Surface		
(15 mph winter, 7.5 mph summer)	0.17	0.25
Roofing	0.33	0.33
Roof Insulation	1.39	1.39
Plywood Deck	0.78	0.78
Air Space[1](framing)	0.85 (6.73)	1.23 (6.73)
Gypsum Wallboard	0.45	0.45
Inside Surface	0.61	0.76
Resistance without insulation		
in air space[2]	4.85	5.47
Resistance with insulation		
(R-value = I) in air space[3]	$3.73 + I$	$3.96 + I$
	------------------	------------------
	$0.936 + .01I$	$0.937 + 0.1I$

Pitched Roof and Ceiling Roof		
Outside Surface (15 mph winter,		
7.5 mph summer)	0.17	0.25
Roofing	0.44	0.44
Building Paper	0.06	0.06
Plywood Deck (5/8)	0.78	0.78
Inside Roof Surface	0.62	0.76
Framing[1]	5.84	5.84
	------	------
Roof resistance[2]	2.24	2.47

[1]Framing is assumed to be 2" × 6", 16" o.d., giving 10 percent of total area.
[2]The formula is:

$$R_{average}^{-1} = R_{framing}^{-1} + 0.9\ R_{other}^{-1}$$

The contribution of open framing to resistance is negligible.
Source: ASHRAE (1977), 22.13-22.22.

Table 2.2 (cont.)

Ceiling and roof resistance

Attic Wall		
Outside Surface	0.17	0.25
Wood Siding	0.87	0.87
Building Paper	0.06	0.06
Sheathing	0.62	0.62
Inside Wall Surface	0.68	0.68
Framing[1]	4.38	4.38
	------	------
Wall Resistance[2]	2.57	2.65
Ceiling		
Upper Surface	0.61	0.76
Insulation (framing)[1]	I (4.48)	I (4.38)
Gypsum Board	0.45	0.45
Inside Surface	0.61	0.76
Ceiling Resistance	$1.67 + I$	$1.97 + I$
	----------------------	----------------------
	$0.9276 + 0.0165I$	$0.9310 + 0.0157I$

[1]Insulation fills the air space.
[2]Framing is assumed to be 2" × 4", 16" o.d., giving 10 percent of total area.
Source: ASHRAE (1977), 22.13-22.22.

Table 2.3

Resistance of windows

Material	Heating R-value	Cooling R-value
Single Glazed (no storm)[1]	.98	1.05
Double Glazed (storm)[1]	2.78	---
Sliding Glass Door, Single Glazed	.88	.94
Sliding Glass Door, Double Glazed	1.32	1.43

[1]Assume wood sash, 80 percent glass.
Source: ASHRAE (1977), 22.24.

Table 2.4

Resistance of floor

Material	Heating R-value	Cooling R-value
Top Surface	.61	.76
Tile	.05	.05
Felt Pad	.06	.06
Plywood	.78	.78
Subfloor	.94	.94
Insulation (R-7)	7.00	7.00
Bottom Surface	.61	.76
Total Resistance[1]	10.05	10.35

[1]The resistance of a slab floor is similar after adjustment for additional insulation. Winter ground temperatures below the frost level are approximately:

$$t_g = 36 + 0.3\, t_e$$

where t_e is design temperature. Basement wall losses due to the temperature gradient above the frost line are neglected. Slab edge losses with insulation are assumed comparable to unheated basement losses. Net heat transfer through the floor in summer is neglected.

Source: ASHRAE (1977, 22.20, Table 4G, and 24.4).

2.2.1. Winter heating load

The combined resistance of ceiling, attic, and roof is calculated as follows: In winter, thermal equilibrium requires:

$$0 = -\frac{A_r}{R_r}(t_a - t_e) - \frac{A_w}{R_w}(t_a - t_e) + \frac{A_c}{R_c}(t_i - t_a) \qquad (4)$$

where t_a is attic temperature, A_c, A_r, A_w are ceiling, roof, and attic wall areas, and R_c, R_r, R_w are resistances. Then:

$$t_i - t_a = \frac{\left(\dfrac{A_r}{R_r} + \dfrac{A_w}{R_w}\right)(t_i - t_e)}{\left[\dfrac{A_c}{R_c} + \dfrac{A_r}{R_r} + \dfrac{A_w}{R_w}\right]} \qquad (5)$$

so the net ceiling heat flux is:

$$\frac{Q}{A_c} = \frac{\left(\dfrac{A_r}{A_c}\dfrac{1}{R_r} + \dfrac{A_w}{A_c}\dfrac{1}{R_w}\right)}{\left(1 + \dfrac{A_r}{A_c}\dfrac{R_c}{R_r} + \dfrac{A_w}{A_c}\dfrac{R_c}{R_w}\right)}(t_i - t_e) = \frac{t_i - t_e}{R_{eff}} \tag{6}$$

For a sample of six representative dwellings with pitched roofs, the average value of A_r/A_c is 1.12 and the average value of A_w/A_c is 0.08. These are used along with the resistances in Table 2.2 to calculate the effective winter ceiling resistance for a dwelling with an attic:

$$R_{eff} = \frac{\left[1 + 1.12\dfrac{R_c}{R_r} + 0.08\dfrac{R_c}{R_w}\right]}{\left[1.12\dfrac{1}{R_r} + 0.08\dfrac{1}{R_w}\right]} = \frac{3.416 + 1.031\,I}{0.9276 + 0.0165\,I} \tag{7}$$

A first order Taylor's expansion of (7) about $I = 3$ gives the approximation:

$$R_{eff} \doteq 3.834 + 0.943\,I \tag{8}$$

For comparison, the resistance of a flat roof and ceiling is approximately:

$$R \doteq 4.064 + 0.960\,I \tag{9}$$

Typical values are:

I	.85	3	6	9
R_{eff}	4.56	6.66	9.35	11.80
approx R_{eff}	4.63	6.66	9.50	12.32
R	4.83	6.95	9.74	12.37
approx R	4.88	6.95	9.83	12.71

On the basis of this comparison, we choose (8) as an adequate approximation of the resistance of all roofs, pitched or flat.

Infiltration is a function of the integrity of the dwelling shell and pressure differentials created by wind, stack effects, and temperature differences. A common method of calculating infiltration effects is to determine the number of air changes per hour in the dwelling K . Then the heat transfer is:

$$\text{Net infiltration heat loss} = 0.018 \, KV \, (t_i - t_e) \qquad (10)$$

where V is the volume of the dwelling (ft^3), 0.018 equals the Btu's required to heat 1 cu. ft. of air by 1 °F, and $t_i - t_e$ is the temperature differential (ASHRAE, 1977. 24.6).

Air changes per hour in most dwellings are in the range $0.5 \leqslant K \leqslant 1.5$ for heating and $1 \leqslant K \leqslant 2$ for cooling. Dwellings with $K < 0.5$ are "stuffy" and $K \geqslant 2$ are "drafty." Experimental measurements by Achenback and Coblentz (1963) give an air change rate:

$$K = 0.25 + 0.02165 \, (\text{wind velocity}) + 0.00833 \, |t_i - t_e| \qquad (11)$$

for an average dwelling. Detailed calculations by Anderson (1973) permit a calculation of the effect of integrity of the shell on this rate. For tight construction, with storm doors and windows, the rate is reduced 14 percent; for loose construction, it is increased 14 percent. Therefore, we multiply the value of K in (11) by a factor:

$$1.14 - 0.28 \, (\text{proportion of window area stormed}) \qquad (12)$$

An additional factor entering thermal calculations is the heat generated internally by occupants and appliances. ASHRAE (1977, 25.17, 25.41) design standards typically assume each occupant generates 225 Btuh in normal activity, while lighting and appliances generate 1200 Btuh. Anderson reports a higher internal load from lighting and appliances of 3083 Btuh, and an effective load-per-occupant of 318 Btuh due to the daily pattern of occupancy. For purposes of calculating design capacity, we use the ASHRAE standards. We follow the usual practice of including internal load in the calculation of air conditioner capacity requirements, but excluding it in heating capacity requirements. The winter heat transfer calculations may now be summarized in Table 2.5.

Table 2.5

Summary of winter heating capacity calculation

Design Btuh is the sum of the following components:

---

1. Wall losses:
(Exterior wall area surrounding heated space, excluding windows)

$$\times \frac{0.9394 + 0.0138 \, I_w}{2.85 + I_w} \cdot (75 - t_e)$$

2. Ceiling losses:

$$\begin{bmatrix} \text{ceiling} \\ \text{area} \end{bmatrix} \cdot (3.834 + 0.943 \, I_c)^{-1} \cdot (75 - t_e)$$

3. Floor losses:

$$\begin{bmatrix} \text{ceiling} \\ \text{area} \end{bmatrix} \cdot [75 - (36 + 0.3 \, t_e)]/10.05$$

4. Window losses:

$$\left(\frac{A_{ws}}{2.78} + \frac{A_{wn}}{0.98} + \frac{A_{sds}}{1.32} + \frac{A_{sdn}}{0.88}\right) \cdot (75 - t_e)$$

5. Infiltration losses:

$$1.14 - \frac{0.28 \, (A_{ws} + A_{sds})}{(A_{ws} + A_{wn} + A_{sds} + A_{sdn})}$$

$$\times \, [0.25 + 0.02165(15) + 0.00833(75 - t_e)]$$

$$\times \, (0.018) \, V \, (75 - t_e)$$

I_w	R-value of wall insulation (0.94 for air gap if no insulation)
I_c	R-value of ceiling insulation
$t_i = 75$	interior design temperature (°F)
t_e	exterior winter design temperature (°F)
A_{ws}	area of stormed windows (ft^2)
A_{wn}	area of non-stormed windows (ft^2)
A_{sds}	area of stormed sliding glass doors (ft^2)
A_{sdn}	area of non-stormed sliding glass doors (ft^2)
V	volume of conditioned space (ft^3)

2.2.2. Summer cooling load

The rate of instantaneous heat gain during the summer is classified by
the mode in which heat enters the residence. Heat gain occurs in the
form of: (1) solar radiation through transparent surfaces; (2) heat conduction through interior partitions, ceilings, and floors; (3) heat conduction through exterior walls and roof; (4) heat generated within the space
by occupants and equipment; (5) energy transfer as a result of ventilation
and infiltration of outdoor air; and (6) all miscellaneous heat gains.

Precise calculation of the effects of solar radiation on air conditioning
requirements necessitates measurement of the angle of incidence of radiation on each surface of the shell, degree of shading, and reflectance of
the surface over the day. Heat flux into the surface satisfies (ASHRAE,
1977, 25.4):

$$(\text{Btuh per ft}^2) = \alpha I_r + (t_o - t_i)/R - \xi\rho \tag{13}$$

where:

t_o outdoor air temperature, °F

α absorptance of surface for solar radiation

I_r solar radiation incident on surface (Btuh/ft^2)

R resistance of surface to radiation and convection heat transfer

ξ emittance of surface

ρ correction for difference between sky and black-body radiation
spectrum

The value of I_r will be a function of latitude, time of day, and the
orientation of the surface. ASHRAE converts this equation to a *sol-air
temperature equivalent*:

$$t_{sa} = t_o + R\,(\alpha I_r - \xi\rho) \tag{14}$$

so that:

$$(\text{Btuh per ft}^2) = (t_{sa} - t_i)/R \tag{15}$$

The temperature t_{sa} is the outdoor temperature that, in the absence of radiation exchanges, gives the same rate of heat entry into the incident surface as exists under standard conditions. The calculation of heat gain combines transient thermal properties of building materials and sol-air equivalent temperatures by the transfer function method (ASHRAE, 1977, 25.27). The rate of instantaneous heat gain will not, however, generally determine instantaneous cooling load. Radiant energy is first absorbed by surfaces that enclose the space. As these surfaces become warmer than the space air, heat is transferred into the room by convection. A transfer function method is used to convert instantaneous heat gain into cooling load.

For hour-by-hour calculation, ASHRAE provides values of the thermal transfer coefficients for roofs and walls under a variety of constructions indexed by weight and average conductivity (ASHRAE, 1977, 25.28, Table 26, and 25.29, Table 27). In Table 2.6 we estimate roof density and weight for the roof materials assumed in Table 2.2. To approximate the conditions maintained in Table 2.2, we examine ASHRAE roofs #22 and #25 with weight of 8 lbs./ft^3 and conductances 0.109 and 0.170 respectively.

Table 2.6

Roof densities and weights

Flat Ceiling and Roof:	Thickness	Density (lbs./ft^3)
Roofing	.375"	70.
Roof Insulation	--	--
Plywood Deck	.625"	34.
Airspace (framing)	5.75"	0.(32)
Wallboard	.5"	50.

Density of section with wood-framing:
   $(.375 \times 70 + .625 \times 34 + 5.75 \times 32 + .5 \times 50)/7.25 = 35.37$

Density of section without framing:
   $(.375 \times 70 + .625 \times 34 + .5 \times 50)/7.25 = 10$

Average density: (assume framing is 10 percent of material)
   $(.10 \times 35.379 + .09 \times 10) = 12.54$ lbs./ft^3

Weight:
Consider a 1 ft^2 section of ceiling. Thickness is (7.25/12) ft., which implies a volume of 0.60417 ft^3. Average weight is 7.8 lbs./ft^2
Density and weight for a pitched roof are 9.06 lbs./ft^3 and 6.47 lbs./ft^2 respectively.

The transfer function method for calculation of instantaneous heat gain through roofs and exterior walls assumes constant indoor temperature and represents outdoor conditions by sol-air equivalent temperatures. Heat gain (at hour h) arising through a roof or wall is:

$$q(h)/A = B(L)t_{sa}(h) - D(L)q(h)/A - t_i C \qquad (16)$$

where:

$$A = \text{indoor surface area of a roof or wall, ft}^2$$

$$q(h) = \text{heat gain, Btuh/ft}^2$$

$$h = \text{solar hour}$$

$$t_{sa}(h) = \text{sol-air temperature at hour } h, \text{°F}$$

$$t_i = \text{indoor temperature, °F}$$

$$B(L), D(L) = \text{lag-polynomials of the transfer function}$$

Table 2.7 presents the transfer function polynomials for ASHRAE roofs #22 and #25. Note that (16) implies:

$$[I + D(L)] \cdot q(h)/A = B(L)t_{sa}(h) - t_i C \qquad (17)$$

In the calculation of $q(h)/A$, initial conditions may be arbitrary provided the polynomial $(I + D(L))$ is invertible. This condition in turn requires that the characteristic equation $F(X) = 1 + d_1 X + d_2 X^2 + d_3 X^3$ have roots that lie within the unit circle in the complex plane. It may be verified that the roofs (and walls) considered in our analysis satisfy this property by direct solution of the cubic equation. The driving function $t_{sa}(h)$ is assumed periodic (with a one-day period) so that calculation of $q(h)/A$ simply requires repeating successive 24-hour cycles in (16) to allow the effect of initial conditions to become negligible.

Table 2.7

Transfer function polynomials for ASHRAE roofs #22 and #25

$B(L) = b_0 + b_1 L + b_2 L^2 + b_3 L^3$

$D(L) = d_0 + d_1 L + d_2 L^2 + d_3 L^3$

Roof #22:

	n=0	n=1	n=2	n=3	U	C
b	0.0012	0.0180	0.0150	0.0011		
d	0.0000	-0.8098	0.1357	-0.0007	0.109	0.0353

Roof #25:

	n=0	n=1	n=2	n=3	U	C
b	0.0043	0.0385	0.0202	0.0007		
d	0.0000	-0.7314	0.1061	-0.0003	0.170	0.0637

U — conductance
C — indoor temperature
L — lag operator

The hourly cycle for $t_{sa}(h)$ will depend on roof and/or wall orientation and the daily cycle of outdoor temperatures. Outdoor temperature will itself follow a pattern determined by the average temperature and daily range of temperatures. Table 2.8 presents the percentages of the daily range used in the calculation of the daily temperature cycle.

Table 2.8

Percentage of the daily range

Time, Hour	Percent	Time, Hour	Percent
1	87	13	11
2	92	14	3
3	96	15	0
4	99	16	3
5	100	17	10
6	98	18	21
7	93	19	34
8	84	20	47
9	71	21	58
10	56	22	68
11	39	23	76
12	23	24	82

Application: temperature at hour h = (maximum temperature) - (percentage) (temperature range) = (daily average temperature) + (0.5-percentage) (temperature range).
    Source: (ASHRAE, 1977, 25.4, Table 3).

ASHRAE (1977, 25.2, Table 2) provides sol-air temperatures for roofs and walls on a day with maximum temperature of 95°F and daily range of temperatures equal to 21°F. We assume that the difference between sol-air temperatures and outdoor temperatures remains constant independent of the daily mean and range of temperatures. Table 2.9 presents the sol-air temperature differences.

Table 2.9

Sol-air temperature differences

Time	Temp. Difference	Time	Temp. Difference
1	0.0	13	25.4
2	0.0	14	28.3
3	0.0	15	30.4
4	0.0	16	29.2
5	0.0	17	26.4
6	17.6	18	17.6
7	26.4	19	0.0
8	29.2	20	0.0
9	30.4	21	0.0
10	28.3	22	0.0
11	25.4	23	0.0
12	23.4	24	0.0

Assume dark-colored surfaces averaged over orientations in the proportions: N—10 percent; S—15 percent; NE, E, SE, SW, W, NW—12.5 percent.
   Source: (ASHRAE, 25.5, Table 2).

To determine cooling load for varied materials and weather conditions, we generate hourly heat flux values when average temperature varies between 70° and 110° (in increments of 5°), daily temperature range varies between 10° and 30° (in increments of 10°), and inside temperature varies between 68° and 84° (in increments of 2°). A convergence criteria for the heat flux profile is suggested by equation (17) evaluated at mean values:

$$[I + D(L)] \cdot \bar{q}/A = B(L) \, \bar{t}_{sa} - t_i \, C \tag{18}$$

which implies:

$$(1 + d_1 + d_2 + d_3)(\bar{q}/A) = (b_0 + b_1 + b_2 + b_3) \, \bar{t}_{sa} - t_i \, C \tag{19}$$

where $\bar{q}$ and $\bar{t}_{sa}$ are average values of $q(h)$ and $t_{sa}(h)$ respectively. From (19) we see that:

$$\bar{q}/A = \frac{(b_0 + b_1 + b_2 + b_3)}{(1 + d_1 + d_2 + d_3)} \bar{t}_{sa} - \frac{C}{(1 + d_1 + d_2 + d_3)} t_i \qquad (20)$$

It is easy to check that the coefficients used in the transfer function method satisfy $C = b_0 + b_1 + b_2 + b_3$ and that conductance U satisfies:

$$U = (b_0 + b_1 + b_2 + b_3)/(1 + d_1 + d_2 + d_3) \qquad (21)$$

Thus convergence of the heat flux profile is accomplished when the sample average of a 24-hour predicted heat flux profile is approximately $U(\bar{t}_{sa} - t_i)$.

Having determined an estimated hourly heat flux profile, we may use the transfer function method to determine hourly cooling load:

$$Q(h)/A = v_0 q(h)/A + v_1 q(h-1)/A - w_1 Q(h-1)/A \qquad (22)$$

where:

$$Q(h) = \text{cooling load at hour } h \text{ (Btuh/ft}^2)$$

$$v_0, v_1, w_1 = \text{coefficients of the room transfer function}$$

The values of v_0, v_1, and w_1 were determined under the assumptions of (1) low room air circulation; (2) 2" wood floor; and (3) frame exterior wall (ASHRAE, 1977, 25.35 - 25.36, Tables 30 and 31). Iterating the stationary heat flux profile until convergence provides hourly cooling loads. Daily cooling load attributable to a surface is then approximately the sum of positive cooling loads arising from that surface over the course of a day. This relationship is only approximate, because the cooling load transfer function applies to the total of all sources of heat flux rather than to the sum of all source cooling loads.

To illustrate the calculation, we present in Table 2.10 the daily profile of outdoor and sol-air temperatures as well as instantaneous heat flux and cooling loads for ASHRAE roof #22 on a day with mean 85°, range 21°, and inside temperature of 75°.

Table 2.10

Output from thermal transfer function calculation for a typical day[1]

Hour	Outside Temp.	Sol-Air Temp. Difference (Sol-Air - outside)	Sol-Air Temp.	Cooling Load Temp.Diff.[2]	Heat Flux
1	77.23	0.00	77.23	13.88	0.91
2	76.18	0.00	76.18	11.65	0.67
3	75.34	0.00	75.34	9.76	0.48
4	74.71	0.00	74.71	8.11	0.32
5	74.50	0.00	74.50	6.69	0.20
6	74.92	17.60	92.52	5.63	0.12
7	75.97	26.40	102.37	6.97	0.41
8	77.86	29.20	107.06	11.09	1.11
9	80.59	30.40	110.99	16.03	1.89
10	83.74	28.30	112.04	20.66	2.59
11	87.31	25.40	112.71	24.50	3.13
12	99.67	23.40	114.07	27.43	3.50
13	93.19	25.40	118.59	29.76	3.78
14	94.87	28.30	123.17	32.14	4.05
15	95.50	30.40	125.90	34.97	4.40
16	94.87	29.20	124.07	37.94	4.76
17	93.40	26.40	119.80	40.27	5.01
18	91.09	17.60	108.69	41.23	5.06
19	88.36	0.00	88.36	39.98	4.77
20	85.63	0.00	85.63	35.39	3.98
21	83.32	0.00	83.32	29.32	3.02
22	81.22	0.00	81.22	24.09	2.24
23	79.54	0.00	79.54	19.93	1.66
24	78.28	0.00	78.28	16.60	1.23

[1]Mean temperature 85°F, temperature range 21°F, inside temperature 75°F, ASHRAE roof #22 with U = 0.109.
[2]Total cooling load temperature difference = 544. Btu/ft^2.

To summarize the relationship of cooling load to standard weather inputs, we have calculated total and maximal cooling load temperature differences for two roofs and four walls over the ranges of temperatures specified above.[2] Each test surface generates 243 observations, which are

[2]A FORTRAN program that implements the thermal transfer function calculations is reproduced in Dubin and McFadden (1983).

used to estimate summary regression formulae. The regression results presented in Table 2.11 are used below in the calculation of daily cooling load and cooling load capacity.

Table 2.11

Summary of regression results

Structure	(Mean-Inside) Temperature	Temperature Range	Constant
Dependent variable is Total Cooling Load Temperature Difference[1]			
1	22.66	-1.032	355.6
2	22.71	-1.012	358.4
3	22.67	-0.9638	362.1
4	22.49	-0.9535	359.5
5	22.58	-0.9526	361.5
6	22.48	-0.9357	360.9
Dependent variable is Maximum Cooling Load Temperature Difference[2]			
1	0.9958	0.2820	25.35
2	1.0010	0.2988	25.55
3	1.0050	0.3196	26.27
4	0.9972	0.3206	26.12
5	1.0020	0.3256	26.30
6	0.9990	0.3300	26.29

[1]Daily cooling load temperature difference = $\alpha_{temp\Delta}$ (mean - inside temperature) + α_{range}(range) + α.

[2]Maximum cooling load temperature difference = $\alpha_{temp\Delta}$ (mean - inside temperature) + α_{range}(range) + α.

Number of observations = 243
All coefficients are significant: R^2's range from 0.9944 to 1.0.

Structure

1. Roof #22, 1" wood, 2" insulation $U = 0.109$
2. Roof #25, 1" wood, 1" insulation $U = 0.170$
3. Exterior frame wall #36, 3" insulation $U = 0.081$
4. Exterior frame wall #37, 2" insulation $U = 0.112$
5. Exterior frame wall #38, 1" insulation $U = 0.178$
6. Exterior frame wall #39, no insulation $U = 0.438$

Noting the similarity in the regression results, we assume that ASHRAE roof #22 and exterior wall #37 provide adequate approximations for cooling load determination. These relationships must be modified for differences in actual levels of thermal resistance.

In the case of a pitched roof, we can calculate an effective cooling resistance for the combined ceiling and roof by using the ceiling and roof resistances given in Table 2.2 and the definition of average heat flux. It is necessary to account for the effective resistance contributed by natural ventilation of the attic. Assume the ASHRAE (1977, 22.23, Table 6, and 24.2) design standard of 0.1 cubic feet per minute per square foot of ceiling area for ventilation, assume the effective cross section of the roof for solar radiation equals the area of ceiling, and neglect the effect of radiation on attic walls. Then thermal equilibrium evaluated at mean values requires:

$$\frac{A_c}{R_r}(\bar{t}_{sa} - t) + (\frac{A_r}{R_r} + \frac{A_w}{R_w})(t - t_a) \tag{23}$$

$$+ (0.018)(.1)(60)A_c(t - t_a) = \frac{A_c}{R_c}(t_a - t_i)$$

where $\bar{t}_{sa}$ is average sol-air temperature, t_a is attic temperature, and t is daily mean temperature. Then:

$$t_a - t_i = \frac{\dfrac{A_c}{R_r}(\bar{t}_{sa} - t) + \left[\dfrac{A_r}{R_r} + \dfrac{A_w}{R_w} + 0.108A_c\right](t - t_i)}{\left[\dfrac{A_c}{R_c} + \dfrac{A_r}{R_r} + \dfrac{A_w}{R_w} + 0.108A_c\right]} \tag{24}$$

and net ceiling heat flux is:

$$\bar{q}/A_c = \frac{\dfrac{1}{R_r}(\bar{t}_{sa} - t) + \left[\dfrac{A_r}{A_c}\dfrac{1}{R_r} + \dfrac{A_w}{A_c}\dfrac{1}{R_w} + 0.108\right](t - t_i)}{1 + \left[\dfrac{A_r}{A_c}\dfrac{1}{R_r} + \dfrac{A_w}{A_c}\dfrac{1}{R_w} + 0.108\right]R_c}$$

$$\tag{25}$$

The values in Table 2.2 and the ratios $A_r/A_c = 1.12$ and $A_w/A_c = 0.08$ imply:

$$\bar{q}/A_c = \left| \frac{0.9310 + 0.0157I}{2.097 + 0.608I} \right| \cdot$$
$$[0.40485(\bar{t}_{sa} - t) + 0.59163(t-t_i)] \tag{26}$$

We may then define:

$$\bar{q}/A_c = U_1^{effective} \cdot (\bar{t}_{sa} - t_i) + U_2^{effective} \cdot (t-t_i) \tag{27}$$

where:

$$U_1^{effective} = \left| \frac{0.3769 + 0.00636I}{2.097 + 0.608I} \right| \tag{28}$$

$$U_2^{effective} = \left| \frac{0.17389 + 0.00293I}{2.097 + 0.608I} \right|$$

From (22) and the values of v_0, v_1, and w_1, we note that average cooling load and average heat flux satisfy $\bar{Q}/A_c = \bar{q}/A_c$. If effective resistance in the attic is approximately uniform over the day, then:

$$\bar{Q}_{attic}/A_c = U_1^{effective} \cdot [\text{\tiny{temperature difference}}^{\text{total cooling load}}] \tag{29}$$
$$+ U_2^{effective} \cdot 24$$

For a flat roof we use the resistance values given in Table 2.2 to obtain:

$$\bar{Q}_{flat}/A_c = \left| \frac{0.937 + 0.1I}{3.96 + I} \right| \cdot [\text{\tiny{temperature difference}}^{\text{total cooling load}}] \tag{30}$$

Typical values of the ceiling cooling load are given in Table 2.12 for $t_i = 75°F$.

Table 2.12

Summer cooling load

Daily Mean	Daily Range	R-Value of Attic Insulation	Pitched[1] Roof Q/A_c	Flat Roof[2] Q/A_c
79.00	12.00	0.00	85.94	102.66
75.00	20.00	0.00	60.20	79.26
84.00	12.00	0.00	116.25	129.47
80.00	20.00	0.00	90.52	106.07
89.00	12.00	0.00	146.57	156.27
85.00	20.00	0.00	120.83	132.87
80.00	30.00	0.00	88.66	103.62
94.00	12.00	0.00	176.88	183.08
90.00	20.00	0.00	151.15	159.68
85.00	30.00	0.00	118.98	130.43
79.00	12.00	10.00	25.76	60.20
75.00	20.00	10.00	18.04	46.48
94.00	12.00	10.00	53.01	107.36
90.00	20.00	10.00	43.30	93.64
85.00	30.00	10.00	35.66	76.49
79.00	12.00	20.00	16.91	53.18
75.00	20.00	20.00	11.84	41.06
94.00	12.00	20.00	34.79	94.85
90.00	20.00	20.00	29.73	82.72
85.00	30.00	20.00	23.40	67.57

[1]Q/A_c is total cooling for 24-hour period in Btu's.
[2]Q/A_c is total cooling for 24-hour period in Btu's.
Internal temperature is 75°F.

For total cooling load temperature difference, we use the regression estimates for roof #22:

$$[\text{\tiny total cooling load / temperature difference}] = 22.66 \cdot (t-t_i) - 1.032 \cdot t_r + 355.6 \qquad (31)$$

We assume the pitched roof formula (29) to be consistent with our assumptions in the heating load calculation.

For design cooling load arising from window gains, sol-air temperature equivalents are given as a function of glazing, orientation, and covering. For windows with draperies, Venetian blinds, or half-drawn roller shades, the formula (ASHRAE, 1977, 25.40, Table 36) is:

$$[\text{Btuh/ft}^2] = -a + bt_e \qquad (32)$$

where $b = 0.8$ for single glazed and $b = 0.6$ for double glazed (stormed) windows, t_e = design temperature, and a has the following values:

Orientation	Single Glazed	Double Glazed	Prop.
N	52	39	.10
NE and NW	33	21	.25
E and W	16	6	.25
SE and SW	24	13	.25
S	43	30	.15
average	30	18	--

The average above is calculated by assuming that square footage of window space in a characteristic dwelling is distributed in the proportions given in the last column.

For the purpose of cooling load calculations, we assume that the window gain effect is essentially uniform over the day so that:

$$[\text{Btuh/ft}^2] = -a + bt \tag{33}$$

The summer heat gain calculation is summarized in Tables 2.13 and 2.14 for daily cooling load and design capacity. The calculations differ in two ways. Wall gains and ceiling gains use total cooling load temperature difference in the daily cooling load calculation and use maximal cooling load temperature difference in the capacity calculation. The second difference concerns the treatment of mean versus design temperatures in the calculation of window and infiltration gains. Cooling load calculations use daily mean temperature under the assumption that relevant gains are uniform due to the thermal flywheel effect. Capacity calculations use design temperature to determine maximal load. Tables 2.15 and 2.16 consider additional allowances for transmission losses in forced hot water and hot air systems.

2.3. Benchmark energy consumption levels

Implicit in the thermal calculations for heating and cooling system design capacities are energy consumption levels under the benchmark behavioral assumptions used. These consumption levels can be

Table 2.13

Summary of summer cooling calculation—daily load

Btu's-per-day is the sum of the following components:

---

1. Wall gains:
(exterior wall area surrounding conditioned space, excluding windows)

$$\times \frac{0.9394 + 0.138 \, I_w}{2.85 + I_w} \times (22.67 \cdot (t - t_i) - 0.9638 \cdot t_r + 362.1)$$

2. Ceiling gains (assume pitched roof):

$$\begin{bmatrix} \text{ceiling} \\ \text{area} \end{bmatrix} \times \frac{(0.3769 + 0.00636 \cdot I_c)}{(2.097 + 0.608 \cdot I_c)} \times (22.66 \cdot (t - t_i) - 1.032 \cdot t_r + 355.6)$$

$$+ \frac{(0.17389 + 0.00293 \cdot I_c)}{(2.097 + 0.608 \cdot I_c)} (t - t_i) \cdot 24$$

3. Window gains (assuming storms removed on windows):

$$(A_{ws} + A_{wn} + A_{sdn}) \cdot (0.8t - 30) \cdot 24 + A_{sds} \cdot (0.6t - 18) \cdot 24$$

4. Internal load (sensible):

$$[1200 + 225 \, (\text{number of occupants})] \cdot 24$$

5. Infiltration gains:

$$24 \cdot 0.018 \cdot V \cdot (t - t_i) \cdot [0.25 + 0.02165 \, (7.5) + 0.00833(t - t_i)]$$

The sum of 1-5 is increased by 25 percent to account for latent heat load (dehumidification) (ASHRAE, 1977, 25.41).

---

t	mean temperature (°F)
t_r	temperature range (°F)
I_w	R-value of wall insulation
I_c	R-value of ceiling insulation
V	Volume of conditioned space

calculated as a function of weather and time to give benchmark HVAC load curves, or can be summed up over the season to give annual HVAC consumption. This section provides the formulae for these calculations.

Consider first the treatment of temperatures through time. When seasonal, monthly, or hourly temperatures are available, they can be used directly in the calculations described below. It is convenient, however,

Table 2.14

Summary of summer cooling calculation—design capacity

Btu's-per-hour at design conditions is the sum of the following components:

1. Wall gains:
(exterior wall area surrounding conditioned space, excluding windows)

$$\times \ \frac{0.9394 + 0.138 \, I_w}{2.85 + I_w} \ \times \ 0.319 \, t_r + 1.0050 \, \Delta t_e + 26.27$$

2. Ceiling gains (assumed pitched roof):

$$\begin{bmatrix} \text{ceiling} \\ \text{area} \end{bmatrix} \ \times \ \frac{(0.3769 + 0.00636 \cdot I_{c)}}{(2.097 + 0.608 \cdot I_{c)}} \ \times \ (0.2820 \, t_r + 0.0058 \, \Delta t_e + 25.35)$$

$$+ \ \frac{(0.17389 + 0.00293 \cdot I_{c)}}{(2.097 + 0.608 \cdot I_{c)}} \cdot \Delta t_e$$

3. Window gains (assuming storms removed on windows):

$$(A_{ws} + A_{wn} + A_{sdn}) \cdot (0.8 t_e - 30) + A_{sds} \cdot (0.6 t_e - 18)$$

4. Internal load (sensible):

$$[1200 + 225 \,(\text{number of occupants})]$$

5. Infiltration gains:

$$0.018 \, V \cdot \Delta t_e \ \cdot \ [0.25 + 0.02165(7.5) + 0.00833 \,(\Delta t_e)]$$

The sum of 1-5 is increased by 25 percent to account for latent heat load (dehumidification) (ASHRAE, 1977, 25.41). $\Delta t_e = t_e - t_i$ where t_e = summer design minimum temperature (°F).

---

for seasonal or annual calculations to use several simple approximations to temperature patterns over time. Let $F(t)$ denote the cumulative distribution function of daily mean temperatures. Then average heating degree-days per day over the year, to base temperature τ, satisfies:

$$HD_\tau = \int_{t_0}^{t_1} \max(0, \ \tau - t) F'(t) \, dt = \int_{t_0}^{\tau} F(t) \, dt \tag{34}$$

where t_0 and t_1 are extreme possible temperatures. Similarly, average cooling degree-days per day to base temperature τ satisfies:

Table 2.15

Hot water system pipe transmission losses

Assume 2.5" black iron pipe with outside diameter of 2.88", 2" insulation with an R-value of 6, delivery temperature 120°, return temperature 80°, basement temperature 40°.

The formula for loss is:

$$\text{Btuh} = \frac{(t_w - t_b)(2\,\pi\,r_s\,L)}{r_s\,(\ln \frac{r_s}{r_o})\,I + R_s}$$

with:

t_w water temperature
t_b basement temperature
r_o outside radius of pipe (ft.) = 1.44/12
r_s outside radius of pipe + insulation (ft.) = 3.44/12
I R-value of insulation (per ft.) = 36
R_s surface resistance = 0.6
L length of pipe

Delivery loss:

$$\text{Btuh} = \frac{80\,(2\pi \cdot 0.287 \cdot L/2)}{0.287 \cdot \ln\left[\frac{3.44}{1.44}\right] \cdot 36 + 0.6} = 7.52L$$

Return loss:

$$\text{Btuh} = \frac{40\,(2\pi \cdot 0.287 \cdot L/2)}{0.287 \cdot \ln\left[\frac{3.44}{1.44}\right] \cdot 36 + 0.6} = 3.76L$$

Total loss: Btuh = 11.28L

Source: ASHRAE (1977) 22.7-22.9, 22.26, and 22.27

$$CD_\tau = \int_{t_0}^{t_1} \max(0,\ t-\tau)F'(t)\ dt = \int_{\tau}^{t_1} [1-F(t)]\ dt \tag{35}$$

Approximate $F(t)$ by a logistic cumulative distribution function:

$$F(t) = (1 + e^{-a-bt})^{-1} \tag{36}$$

Table 2.16

Heating duct transmission losses

Assume 600 fpm velocity, 4" x 10" ducts, 2" insulation with R-value of 6, average air temperature in delivery duct 120°, basement temperature 40°.

The formula for loss is:

Btuh $= PL \ (t_d - t_b)/I$

where:

P perimeter (ft.)
L length (ft.)
t_d average duct temperature (°F)
t_b basement temperature (°F)
I R-value of insulation

Assuming 80 percent of ducting is for delivery and neglecting return heat loss, the total loss is:

Btuh $= (2.33)(.8) \ L \ (80)/6 = 24.9L$

Source: ASHRAE (1977), Chap. 30.

Then:

$$HD_\tau = \frac{1}{b} \ \ln \ (1 + e^{a+b\tau}) \tag{37}$$

$$CD_\tau = \frac{1}{b} \ \ln \ (1 + e^{-a-b\tau}) \tag{38}$$

If HD_{65} and CD_{65} are given, then the parameters a and b can be determined by solving:

$$1 = e^{-b \cdot CD_{65}} + e^{-b \cdot HD_{65}} \tag{39}$$

$$a = b(HD_{65} - CD_{65} - 65) \tag{40}$$

Then HD_τ and CD_τ can be calculated for other bases. The value of b is quickly calculated by iteratively splitting the interval containing the solution, starting from:

$$\ln 2 \, / \, \max(HD_{65}, CD_{65}) \; \leqslant \; b \; \leqslant \; \ln 2 \, / \, \min(CD_{65}, HD_{65}) \tag{41}$$

Note that $F(t)$ has mean $-a/b$, variance $\pi^2/3b^2$, and a 95 percent temperature range $t_{high} - t_{low} = 2.9444/b$. For the NIECS data, national average values are $b = .1218$ and $a = -6.870$, implying annual mean temperature 56.4, standard deviation 14.9, and 95 percent temperature range 24.2, or $32.2 \leqslant t \leqslant 80.6$. These match the actual distribution of mean daily temperatures for average U.S. locations quite well.

2.3.1. Heating load calculation

Space heat capacity as a function of ambient temperature and thermostat setting may be interpreted as a measure of average hourly consumption of delivered energy over a day with the specified temperatures. Therefore, benchmark consumption levels can be calculated from the capacity models by replacing design temperatures with the seasonal pattern of daily mean temperatures. Delivered energy per hour on a winter day with mean ambient temperature t and thermostat setting τ is, from Table 2.6:

$$Q = [A_w U_w + A_c U_c + A_{win} U_{win}](\tau - t) + A_c U_f(\tau - t_g) \tag{42}$$
$$+ \, \theta V[.01035 + .00015 \, (\tau - t)](\tau - t) - \text{INTERNAL}$$

The notation is:

A_w, A_c, A_{win} wall, ceiling, and window areas
U_w, U_c, U_{win}, U_f average conductivities of wall, ceiling, and floor
θ window infiltration loss factor
V volume
t_g ground temperature, *assumed constant throughout the winter*
INTERNAL   internal load from occupants and appliances

In this formula, no attempt is made to correct for the effect of the non-linearity in infiltration with temperature difference over the daily temperature cycle. For typical daily temperature ranges, this correction is negligible at the level of precision of the overall calculation. Rewrite (42) in the form:

$$Q = w_0 + w_1(\tau - t) + w_2(\tau - t)^2 \tag{43}$$

with:

$$w_0 = A_c U_f(\tau - t_g) - \text{INTERNAL}$$

$$w_1 = A_w U_w + A_c U_c + A_{win} U_{win} + .010350 V$$

$$w_2 = .000150 V$$

Then the annual average delivered heat (Btuh) is given by:

$$Q_{seas} = \int_{t_0}^{\tau} \max(Q(t), 0) F'(t) \, dt \tag{44}$$

If $w_0 < 0$, then there is a balance temperature $t_b < \tau$ above which heat is not required:

$$t_b = \tau + \frac{w_1}{2w_2} [1 - (1 - 4w_2 w_1^{-2} \min(w_0, 0))^{1/2}]$$

Then (44) can be rewritten:

$$Q_{seas} = \int_{t_0}^{t_b} \left[w_0' + w_1'(t_b-t) + w_2'(t_b-t)^2 \right] F'(t) \, dt$$

$$= w_0'F(t_b) + w_1' \int_{t_0}^{t_b} F(t) \, dt + 2w_2' \int_{t_0}^{t_b} (t_b-t)F(t) \, dt$$

where:

$$w_0' = w_0 + w_1(\tau-t_b) + w_2(\tau-t_b)^2 = 0$$

$$w_1' = w_1 + 2w_2(\tau-t_b)$$

$$w_2' = w_2$$

Using integration by parts,

$$\int_{t_0}^{t_b} (t_b-t)F'(t) \, dt = \int_{t_0}^{t_b} F(t) \, dt$$

$$\int_{t_0}^{t_b} (t_b-t)^2 F'(t) \, dt = 2 \int_{t_0}^{t_b} (t_b-t)F(t) \, dt$$

Using the approximation (36) to the seasonal temperature distribution yields:

$$F(\tau) = (1 + e^{-a-b\tau})^{-1} \tag{45}$$

$$\int_{t_0}^{\tau} F(t) \, dt = \frac{1}{b} \ln (1 + e^{a+b\tau}) \tag{46}$$

$$2 \int_{t_0}^{\tau} (\tau-t)F(t) \, dt = \frac{2}{b} \int_{t_0}^{\tau} \ln (1 + e^{a+bt}) \, dt \tag{47}$$

$$= 2 \, \gamma(a+b\tau)/b^2$$

where:

$$\gamma(\lambda) = \int_{-\infty}^{\lambda} \ln(1 + e^s) \, ds = \int_0^{e^\lambda} \ln(1 + x) \, \frac{dx}{x} \tag{48}$$

Note that for $0 \leqslant x \leqslant 1$:

$$\ln(1 + x) = \sum_{k=1}^{\infty} \frac{(-1)^{k-1}}{k} x^k \tag{49}$$

Also, with error at most 10^{-5} for $0 \leqslant x \leqslant 1$:

$$\frac{\ln(1 + x)}{x} \doteq a_1 + a_2 x + a_3 x^2 + a_4 x^3 + a_5 x^4 \tag{50}$$

with:

$a_1 = .99949556 \qquad a_4 = -.13606275$
$a_2 = .49190896 \qquad a_5 = .03215845$
$a_3 = .28947478$

Then:

$$\gamma(0) = \int_0^1 \ln(1 + x) \, \frac{dx}{x} = \sum_{k=1}^{\infty} \frac{(-1)^{k-1}}{k} \int_0^1 x^{k-1} \, dx \tag{51}$$

$$= \sum_{k=1}^{\infty} (-1)^{k-1}/k^2 = \pi^2/12$$

and for $\lambda < 0$, with error at most 10^{-5}:

$$\gamma(\lambda) = \int_0^1 \ln(1 + x) \, \frac{dx}{x} - \int_{e^\lambda}^0 \ln(1 + x) \, \frac{dx}{x} \tag{52}$$

$$= \frac{\pi^2}{12} - \sum_{k=1}^{5} a_k \int_{e^\lambda}^{1} x^{k-1} \, dx$$

$$= \frac{\pi^2}{12} + \sum_{k=1}^{5} \frac{a_k}{k} (e^{k\lambda} - 1)$$

For $\lambda > 0$, with error at most 10^{-5}:

$$\gamma(\lambda) = \int_{-\infty}^{0} \ln(1 + e^s) \, ds + \int_{0}^{\lambda} \ln(1 + e^s) \, ds \tag{53}$$

$$= \frac{\pi^2}{12} + \int_{0}^{\lambda} s \, ds + \int_{0}^{\lambda} \ln(1 + e^{-s}) \, ds$$

$$= \frac{\pi^2}{12} + \frac{\lambda^2}{2} + \int_{e^{-\lambda}}^{1} \ln(1 + x) \frac{dx}{x}$$

$$\doteq \frac{\pi^2}{12} + \frac{\lambda^2}{2} + \sum_{k=1}^{5} \frac{a_k}{k} (1 - e^{-k\lambda})$$

Defining $\alpha = \max(\lambda, 0)$, $\beta = e^{-\alpha}$, $\delta = e^{\lambda}$, and $c_k = a_k/k$, all cases can be combined in the formula:

$$\gamma(\lambda) \doteq \frac{\pi^2}{12} + \frac{\alpha^2}{2} + \sum_{k=1}^{5} c_k \beta^k (\delta^k - 1) \tag{54}$$

We summarize the annual average delivered heat per hour as:

$$Q_{seas} = w_0/(1 + e^{-a-b\tau}) + (w_1/b) \ln(1 + e^{a+b\tau})$$
$$+ (2w_2/b^2) \gamma(a+b\tau) \qquad\qquad \text{for } w_0 \geqslant 0$$

$$Q_{seas} = (w_1'/b) \ln(1 + e^{a+bt_b})$$

$$+ (2w_2'/b^2)\, \gamma(a+bt_b) \qquad\qquad\qquad \text{for } w_0 < 0$$

2.3.2. Cooling load calculation

Delivered energy per hour on a summer day with mean ambient temperature t and thermostat setting τ is, from Table 2.13:[3]

$$Q = S_0 + S_1(t - \tau) + S_2(t - \tau)^2$$

where:

$$S_0 = [\, (A_w U_w (362.1 - 0.9638 \cdot t_r)$$

$$+ \; A_c U_1^{\mathit{eff}}(355.6 - 1.032 \cdot t_r))/24$$

$$+ \; (A_{ws} + A_{wn} + A_{sdn})(0.8\tau - 30)$$

$$+ \; A_{sds}(0.6\tau - 18) + \text{INTERNAL} \,]\cdot 1.25$$

$$S_1 = [\, (A_w U_w \cdot 22.67 + A_c U_1^{\mathit{eff}} \cdot 22.66$$

$$+ \; A_c U_2^{\mathit{eff}} \cdot 24)/24 + (A_{ws} + A_{wn} + A_{sdn})(0.8)$$

$$+ \; A_{sds}(0.6) + 0.00742 \cdot V \,]\cdot 1.25$$

$$S_2 = (0.00015 \cdot V) \cdot 1.25$$

[3]Notation is given in Tables 2.13 and 2.14.

Then annual average delivered cooling (Btuh) is given by:

$$Q_{seas}^{AC} = \int_{\tau}^{t_1} \max(Q(t), 0)F'(t) \, dt \tag{55}$$

If $S_0 < 0$, then there is a balance temperature $t_b > \tau$ below which cooling is not required:

$$t_b = \tau - \frac{S_1}{2S_2}[1 - (1-4S_2S_1^{-2}\min(S_0, 0))^{1/2}]$$

Then (55) can be rewritten:

$$Q_{seas}^{AC} = \int_{t_b}^{t_1} [S_0' + S_1'(t - t_b) + S_2'(t - t_b)^2]F'(t) \, dt \tag{56}$$

where:

$$S_0' = S_0 + S_1(t_b - \tau) + S_2(t_b - \tau)^2 = 0$$

$$S_1' = S_1 + 2S_2(t_b - \tau)$$

$$S_2' = S_2$$

Note that we may relate the integral in (56) to the form evaluated in the heating load calculation because:

$$\int_{t_b}^{t_1} [S_0' + S_1'(t - t_b) + S_2'(t - t_b)^2]F'(t) \, dt$$

$$= \int_{t_0}^{t_1} [S_0' - S_2'(t_b - t) + S_2'(t_b - t)^2]F'(t) \, dt$$

$$- \int_{t_0}^{t_b} [S_0' - S_1'(t_b - t) + S_2'(t_b - t)^2]F'(t) \, dt$$

$$= S'_0 - S'_1(t_b - \mu) + S'_2[(t_b - \mu)^2 + var(t)]$$

$$- \int_{t_0}^{t_b} [S'_0 - S'_1(t_b - t) + S'_2(t_b - t)^2]F'(t)\, dt$$

where $\mu = -a/b$ and $var(t) = \pi^2/3b^2$. The cases $S_0 < 0$ and $S_0 \geqslant 0$ imply:

$$S'_0 = \begin{cases} 0 & S_0 < 0 \\ S_0 & S_0 \geqslant 0 \end{cases}$$

$$S'_1 = \begin{cases} S_1 + 2S_2(t_b - \tau) & S_0 < 0 \\ S_1 & S_0 \geqslant 0 \end{cases}$$

$$S'_2 = S_2$$

Duct losses for air conditioning are ignored, so the expression for Q_{seas}^{AC} gives gross air conditioner output. For duct and pipe systems, additional furnace output is required to offset transmission losses. These losses can be divided into a component due to conduction losses from the delivery system and a component due to heat gains and losses of the delivery system under cyclic operation. The first component is to a close approximation proportional to heat delivered, and the coefficient of proportionality can be obtained from the calculation of capacity requirements for non-central and central systems. Thus:

$$Q_D = Q_{seas} \cdot \text{SHEATD/SHEATN} \tag{57}$$

$$Q_P = Q_{seas} \cdot \text{SHEATP/SHEATN} \tag{58}$$

where SHEATN, SHEATD, and SHEATP are capacities of non-central, duct, and pipe systems respectively, and Q_D and Q_P are seasonal furnace outputs net of cyclic losses.

Seasonal efficiencies of heating equipment depend on climate, through cyclic heat loss. Empirical seasonal efficiencies of heating equipment, or *coefficients of performance*, can be obtained from ASHRAE. Distribution cyclic heat losses are small relative to the furnace losses and will be ignored. The coefficients for gas, oil, electric resistance (baseboard), and heat pump are respectively:

$$COP_G = .46 + .0146 \cdot HD_{65} \tag{59}$$

$$COP_O = .404 + .0130 \cdot HD_{65} \tag{60}$$

$$COP_E = 1.0 \tag{61}$$

$$COP_{HP} = 1.94 + \frac{2.85}{HD_{65}} + \frac{.96}{CD_{65}} - .046\ HD_{65} - .081\ CD_{65} \tag{62}$$

The efficiency loss in central electric resistance units is relatively small, and is ignored. For air conditioning, the coefficient of performance is approximately:

$$COP_{AC} = 3.44 + \frac{.744}{HD_{65}} + \frac{1.23}{CD_{65}} - .036\ HD_{65} - .038\ CD_{65} \tag{63}$$

The approximations in equations (62) and (63) are discussed in the Appendix to this chapter.

The base technological calculations of seasonal energy consumption can now be summarized. Take, for example, a gas-fired forced-air system. Energy input in MBH over the year is $8.76\ Q_D/COP_G$. Similarly, an electric baseboard system requires an input of $8.76\ Q_{seas}/COP_E$, while an air conditioner requires $8.76\ Q_{seas}^{AC}/COP_{AC}$. Multiplied by marginal fuel prices, these figures give the technologically based operating costs of alternative systems.

These calculations are carried out for specified winter and summer thermostat settings. Repeating the calculations for a 1-degree change in the thermostat setting and taking differences yield an overall calculation

of the seasonal price of comfort. In carrying out these "price" calculations, we ignore the very small change in w_0 induced by the thermostat
change.

The seasonal calculations just completed can also be applied to time
periods within a season, such as billing periods. The temperature distribution $F(t)$ should then be that applicable for the period in question.
The logistic approximation used for the seasonal temperature distribution requires some modification for use in billing periods.

A more accurate temperature distribution can be obtained using
degree-day calculations for alternative bases. Let H_υ, C_υ denote heating
and cooling degree-days (per day) to base υ; then:

$$H_\upsilon = \frac{1}{b} \ln (1+e^{b(\upsilon-\mu)}) = \upsilon-\mu + \frac{1}{b} \ln (1+e^{-b(\upsilon-\mu)})$$

$$= \max(0, \upsilon-\mu) + \frac{1}{b} \ln (1+e^{-b|\upsilon-\mu|})$$

$$C_\upsilon = \int_\upsilon^\infty (t-\upsilon)F'(t) \, dt = \frac{1}{b} \ln (1+e^{-b(\upsilon-\mu)})$$

$$= \max(0, \mu-\upsilon) + \frac{1}{b} \ln (1+e^{-b|\upsilon-\mu|})$$

Note that $C_\upsilon-H_\upsilon = \int_{-\infty}^{+\infty} (t-\upsilon)F'(t) \, dt = \mu-\upsilon$. For a base $\tau \geqslant \upsilon$, one
has:

$$(\tau-\upsilon)F(\upsilon) \leqslant H_\tau-H_\upsilon = \int_\upsilon^\tau F(t) \, dt \leqslant (\tau-\upsilon)F(\tau)$$

$$(\tau-\upsilon)[1-F(\tau)] \leqslant C_\upsilon-C_\tau = \int_\upsilon^\tau [1-F(t)] \, dt \leqslant (\tau-\upsilon)[1-F(\upsilon)]$$

Given C_υ, H_τ for $\tau \geqslant \upsilon$, consider the function:

$$G(b) = (1 - e^{-bH_\tau}) \cdot (1 - e^{-bC_\upsilon}) \cdot e^{b(\upsilon - \tau + H_\tau + C_\upsilon)} - 1$$

derived by eliminating μ from the equations for H_τ and C_υ. This function has $G(0) < 0$ and $G'(0) > 0$. If $G(1) > 1$, then a unique solution can be obtained by successive interpolation, with the consistent value:

$$\mu = \upsilon + C_\upsilon + \frac{1}{b} \ln (1 - e^{-bC_\upsilon})$$

On the other hand, if $G(1) < 1$, as will be the case if C_υ or H_τ is sufficiently small, then the temperature distribution has little mass in the range of υ and τ, where balance temperatures are attained. Outside this range, marginal heating and cooling requirements are linear in the temperature differential, except for small stack and ground effects in heating, which can be neglected. Hence, in this case it is a good approximation to assume that mean temperature for the period under study is concentrated at $\mu = \upsilon + C_\upsilon - H_\upsilon$, and set $Q_{AC} = Q(\mu)$, with an analogous procedure for heating.

2.4. Characteristics of single-family dwellings

The thermal calculations in the preceding sections require information on wall, window, and ceiling areas, volume, and feet of pipe or ducting for central heating systems. The NIECS/PNW data do not provide this level of detail, but do provide (incomplete) information on square footage and numbers of rooms, floors, and windows. To fill this gap, we have sampled seven typical dwellings, and from their detailed characteristics obtained relationships between the required variables and those observed in NIECS/PNW. Table 2.17 lists the measured characteristics.

A series of regressions on these seven observations provide a link from variables in NIECS/PNW to structural characteristics, as follows:

1) ln (wall area including windows) = 2.96 + 0.92 ln (no. floors) + 0.57 ln (ceiling area) ; $R^2 = 0.99$ $\sigma^2 = 19$

2) ln (ceiling area) = -0.04 + 0.815 ln (no. floors) + 1.006 ln (house sq. ft.) ; $R^2 = 0.996$ $\sigma^2 = 0.007$

Table 2.17

Characteristics of typical dwellings

Variable	Dwelling						
	1	2	3	4	5	6	7
Floors	1	2	1	1	2	1	2
Rooms	5	9	4	5	9	9	8
Baths	1	2	1	1	3	3	2.5
Bedrooms	3	7	2	3	4	4	4
Sq. ft.	768	1404	576	1024	3128	2552	1848
Sq. ft. largest room	315	275	153	227	271	433	293
Ceiling area	768	864	576	1024	1888	2552	924
Roof area	810	1222	607	1109	1990	2552	974
Attic wall area	96	81	96	0	171	0	0
Wall area excl. windows	851	1396	736	931	2472	1532	1552
Number picture windows	0	0	0	0	6	0	1
Sq.ft. picture windows	0	0	0	0	248	0	24
Number sliding glass doors	0	0	0	0	6	2	2
Sq.ft. sliding glass doors	0	0	0	0	308	140	77
Number other windows	10	16	8	11	7	13	20
Sq.ft. other windows	45	60	32	94	66	128	314
Volume	6144	10656	4608	8192	27232	22491	15246
Hot air system:							
registers	11	15	9	9	21	20	20
ft. duct	92	114	81	104	292	251	226
Hot water system:							
radiator	6	10	5	5	12	14	14
ft. pipe	160	230	128	144	564	530	344

3) $\ln$ (volume) $= 2.19 + 0.80 \ln$ (no. floors) $+ 0.98 \ln$ (ceiling area) ;
      $\bar{R}^2 = 0.98$ $\sigma^2 = 0.01$

Average area per picture window $= 38.9$; average area per other window $= 8.7$; average area per sliding glass door $= 52.5$.

4)   (average roof area / ceiling area) $= 1.12$; $\sigma^2 = 0.15$

5)   (average attic wall area / ceiling area)= 0.08 ; $\sigma^2 = 0.07$

6)   number of registers = 2.55 + 1.07 (rooms) + .003 (house square feet) ; $\bar{R}^2 = 0.97$ $\sigma^2 = 4.52$

7)   feet of duct = 3.89 (number of registers) + 0.067 (house square feet) ; $\bar{R}^2 = 0.99$ $\sigma^2 = 427$

8)   ln (feet of hot water pipe) = -1.95 + 1.03 ln (house square feet) ; $\bar{R}^2 = 0.99$ $\sigma^2 = 0.02$

9)   number of radiators = 1.04 (number of rooms) + 0.0014 (house square feet) ; $\bar{R}^2 = 364$ $\sigma^2 = 0.96$

We use these equations to estimate structural characteristics of the NIECS/PNW dwellings, except that for hot water systems, we assume baseboard radiators rather than conventional radiators, and use the ASHRAE design standard that 1 linear foot of baseboard radiator is required per 645 Btuh designed capacity of the heating system.

The proportion of window area to total wall area in the typical houses ranges from 0.04 to 0.27. We shall assume that for the NIECS/PNW houses, this proportion is bounded between 0.03 and 0.7, and use these bounds if the regression predicts a more extreme value.

The NIECS/PNW data report square footage of the dwelling as estimated by the respondent. There is evidence in the NIECS data, however, that these responses are subject to error. Therefore, we regress reported square footage on several variables that we believe to be measured more accurately, and use the predicted values from this equation in our analysis. The method was to remove the accurately measured square footage of the largest room from the reported total square feet, predict the square footage of the remainder of the dwelling, and then add back in the largest room square footage. The estimated equation is given in Table 2.18. Some average characteristics of the NIECS/PNW dwellings are given in Table 2.19.

Typical house 7 is taken from ASHRAE (1977, 24.7-24.9), which calculates its heating system capacity to be 114 MBtuh for a Syracuse, N.Y. location with a design temperature of -10°F when there is no wall or attic insulation. The thermal program developed here, using the same inputs as are provided for the NIECS/PNW households, and the design temperature for this location, yields a capacity of 131 MBtuh. The thermal program yields a central air conditioning capacity of 65 MBtuh for this

Table 2.18

Regression of log square feet per room (for rooms other than
the largest) for NIECS households

Variable	Region 1	Region 2	Region 3	Region 4
Baths	.149	.118	.184	.124
	(.071)	(.045)	(.043)	(.053)
Floors	.011	.055	.067	.131
	(.062)	(.037)	(.048)	(.052)
Income (000)	.0039	.0039	.0036	.0032
	(.0033)	(.0019)	(.0020)	(.0021)
Year Built	.0020	.0026	.0002	.0037
(1930-1978)	(.0024)	(.0014)	(.0015)	(.0018)
Largest Room	-.081	-.002	-.15	-.07
L-shaped	(.143)	(.085)	(.07)	(.07)
No. Doors	-.007	.032	.029	.019
	(.036)	(.027)	(.022)	(.025)
No. Windows	.0043	.0058	.0025	.0060
	(.0057)	(.0033)	(.0041)	(.0036)
Log No. Rooms	-.377	-.317	-.592	-.577
	(.171)	(.103)	(.101)	(.112)
Heating Degree-Days	.078	-.011	-.059	.010
(1000)	(.082)	(.038)	(.060)	(.018)
Cooling Degree-Days	-.134	.247	-.035	-.056
(1000)	(.291)	(.109)	(.098)	(.069)
Value of House	.0015	.0020	.0016	.0018
(1000)	(.0012)	(.0005)	(.0009)	(.0008)
SMSA Dummy	.034	-.46	-.16	-.04
	(.074)	(.49)	(.05)	(.08)
Urban Area Dummy	-.091	.19	.078	-.027
	(.068)	(.49)	(.042)	(.080)
Constant	5.30	4.88	6.06	5.68
	(.70)	(.35)	(.44)	(.25)
$\bar{R}^2$	.16	.17	.17	.22
Observations	230	494	432	253

house under the Syracuse summer design temperature of 90 with a daily
range of 20. The corresponding ASHRAE calculation using actual
characteristics of the shell gives an air conditioning capacity of 44
MBtuh.

Table 2.19

Selected characteristics of NIECS/PNW households

Variable	NIECS Mean (estimated)	Typical House Mean	PNW Mean (estimated)
Floors	1.42	1.43	1.29
Rooms	6.06	7.00	5.97
Baths	1.49	1.93	1.17
Square feet	1572 (1553)[1]	1614	1513
Volume	11400	13510	11930
No. Windows	13.0	14.6	10.43
Window Area	(179.4)	219.4	(246.9)
Wall Area Exc. Windows	(1506)	1353	(1341)
Ceiling Area	(1175)	1228	(1292)
Feet Duct	(151.0)	166	(156.1)
Feet Pipe	(265.6)	300	(273.9)
Space Heat Cap. Net of Distribution Losses, MBtuh	(45.5)	---	(46.7)
Central AC Capacity, MBtuh	(34.71)	---	(28.6)
Proportion with Attic Insul. Average R-value Attic	81.9 (17.41)	---	81.0 (20.28)
Average R-value Wall	(7.03)	---	(8.93)

[1]Correlation .94 between observed and estimated.

The usage calculations in Section 2.3 applied to the NIECS and PNW households imply the coefficients of performance and usages in Table 2.20. Note that these are averages over all dwellings of the performance of the specified equipment *if* it were installed in every dwelling, *not* the performance of equipment actually installed.

To test the sensitivity of the thermal model, we have calculated capacity and usage under two alternative levels of building thermal characteristics. The first alternative is an uninsulated dwelling without storm windows or double glazing. The second alternative is the ASHRAE 90-75 voluntary thermal standard for new construction. Under this standard, all windows are stormed or double glazed, walls and ceiling are insulated, heating and cooling system capacities are reduced, and tight construction is used to reduce infiltration. The ASHRAE standards vary by region, as presented in Table 2.21. Table 2.22 summarizes the

Table 2.20

Energy usage characteristics of NIECS/PNW dwellings

	NIECS		PNW	
	Mean	SD	Mean	SD
Coefficients of Performance				
Gas	.64	.08	.70	.05
Oil	.57	.07	.62	.05
Heat Pump	2.3	2.4	3.2	2.7
Air Conditioner	4.0	3.1		
Energy Consumption[1]				
Electric resistance	99420	51800	130800	60610
Gas	151300	71310	186700	87070
Oil	171600	80800	211600	98700
Heat Pump	54600	32800	47960	27390
Air Conditioner	6976	5780	1200	1234
Energy Price of Comfort[2]				
Electric Resistance	4726	2120	5351	2614
Gas	7446	3455	7723	3978
Oil	8447	3922	8753	4511
Heat Pump	2523	1249	1918	1046
Air Conditioner	684	540	155	130

[1]In 10^3 Btu's, net of distribution losses.
[2]In 10^3 Btu's, per degree thermostat setting, net of distribution losses.

differences in capacities and energy consumption under these alternatives, for dwellings built since 1970.

Note that observed thermal performance achieves a substantial fraction of that achievable under the ASHRAE 90-75 standards. In electric resistance heating, for example, 81 percent of the potential conservation is achieved. Substantial conservation is still attainable, however, from the ASHRAE standard: for electric resistance heating, electricity consumption could have been reduced 29 percent relative to actual construction, with comparable reductions for other heating systems. Table 2.23 gives the sample average capital costs (in 1981 prices) of the thermal improvements and heating system for the observed construction since 1970 and for the ASHRAE standards.

These costs are taken from an equipment and construction costing program described in Cowing-Dubin-McFadden (1982). Note first that for electric baseboard resistance heating and heat pumps, the savings in equipment cost from reduced design capacity requirements and downsizing more than offset the added cost of meeting the ASHRAE

Table 2.21

Thermal characteristics by region

	Region 1 Northeast	Region 2 North Central	Region 3 South	Region 4 West
R-value ceiling insulation	17.14	17.14	19.5	19.5
R-value wall insulation	15.44	15.44	9.45	9.45
Reduction in heating design temp. differential	12	14	12	14
Reduction in cooling design temp. differential	7	6	6	5

standards, even before the reduction in life-cycle costs from reduced operating cost is taken into account. On the other hand, for gas forced-air systems, there is an average increment in capital cost of $486.70 required to meet the ASHRAE standards and reduce energy consumption by 28,790,000 Btu's/year. At an average gas price of $3.54 per 10^6 Btu's, the operating cost savings is approximately $102/year.

Ignoring the effects of finite dwelling and equipment life, the real rate of return to adoption of the ASHRAE standards is 21 percent. Because this rate exceeds the real interest rate for most consumers who are free of credit constraints, it appears that improvement of thermal performance to meet the ASHRAE standard should in fact benefit most consumers and may be adopted voluntarily if consumers are fully appraised of the life-cycle costs. This conclusion is subject to a caveat that a comparison is being made between actual and standard thermal levels for gas forced-air heat, irrespective of the type of heat actually chosen. In fact, actual insulation levels are higher for electrically heated homes than for homes heated by other fuels, as should be expected when thermal performance is adjusted to minimize life-cycle cost. This will tend to lead the preceding calculation to overstate the benefit attainable from imposing standards on electrically heated homes, and understate the benefit for homes using other fuels. A more careful behavioral analysis of joint choice of heating fuel and thermal shell performance will be reported separately.

We conclude this chapter with a few comments on the uses and limitations of the thermal and costing models we have developed. First, it is *not* our objective to construct a detailed thermal model suitable for

Table 2.22

Sensitivity of HVAC system to thermal characteristics (NIECS)[1]

	Observed Dwelling	Uninsulated Dwelling	ASHRAE 90-75 Standards
Air Conditioning			
Capacity[2]	34.08	57.56	23.53
Energy Consumption (1000 Btu's)	6883	10060	5590
Electric Resistance Heat			
Capacity[3]	47.20	85.02	29.63
Energy Consumption[4]	107500	195200	86920
Gas Forced-Air			
Capacity[2]	51.86	89.28	33.89
Energy Consumption[4]	106400	287000	128300
Oil Forced-Air			
Capacity[2]	51.86	89.28	33.89
Energy Consumption[4]	181190	325400	145500
Heat Pump			
Capacity[2]	51.86	89.28	33.89
Energy Consumption[4]	57710	104600	46560

[1]Houses built after 1970.
[2]Capacity for forced-air central system in MBtuh.
[3]Capacity of non-central baseboard system in MBtuh.
[4]Annual energy consumption in 10^3 Btu's, including distribution losses.

Table 2.23

Costs of thermal improvements

	Observed Dwellings	ASHRAE Standard
(1) Insulation cost	790.30	1,136.00
(2) Storm/Double glazing cost	263.20	444.60
(3) Electric resistance capital cost	1,942.40	1,386.00
(4) Total (1) + (2) + (3)	2,995.90	2,966.60
(5) Gas forced-air capital cost	2,392.30	2,351.90
(6) Total (1) + (2) + (5)	3,445.80	3,932.50
(7) Heat pump capital cost	9,293.00	5,274.00
(8) Total (1) + (2) + (7)	10,346.50	6,854.60

engineering new dwellings or carrying out energy audits for existing structures, and it would be a mistake to try to use the model for these purposes. The data requirements for such modeling are greater by an order of magnitude than the structural information in the NIECS or

PNW data sets. Second, it *is* our objective to utilize the data available from NIECS/PNW to approximate thermal requirements across a statistical sample in a way that explained most of the technologically determined scale of capacity and usage. The outputs of the thermal model can then be used as inputs to an analysis of choice behavior, with econometric models explaining behavioral deviations from the engineering base.

The thermal and costing models we have developed appear to give a much more satisfactory basis for pricing out alternative HVAC systems than one could achieve using simple formulae for cost per square foot or cost per square foot degree-day. Further, the implied energy consumption under alternative weather conditions should be adequate for indexing the expected operating costs of alternative systems.

We see several advantages to combining the simple engineering thermal model we have developed and a behavioral analysis of consumer response. We can avoid the problems of a pure econometric approach that "burns degrees of freedom" to explain usage variations that are technically determined. We also avoid a pure engineering model that fails to account for economic behavioral response. In addition, the use of the thermal model as an input to the behavioral analysis allows one to calculate readily the technical *and* behavioral response of households to energy policies. This permits a logically consistent and complete method for translating policy affecting voluntary or mandatory building standards into technical consequences in terms of capital cost and energy requirements, modified by consumers' behavioral responses to these consequences.

2.5. Appendix: Seasonal heating and cooling efficiencies of air conditioners and heat pumps

The coefficient of performance of air conditioners and heat pumps in the cooling mode from ASHRAE graphs is approximately:

$$COP = 1/(.235 + .0051\ (t-\tau))$$

where t is daily mean ambient temperatures and τ is thermostat setting. As a basis for a seasonal efficiency calculation, consider a typical residence in which the average cooling load over a day in the cooling season (defined by $t > \tau = 75$) is 17900 + 791 (t-75) Btuh. The energy output per hour for cooling, averaged over the year, is:

$$Q_{out} = \int_{75}^{t_1} [17900 + 791\ (t-75)]F'(t)\ dt$$

where *F(t)* is the distribution of daily mean temperatures. The corresponding energy input is:

$$Q_{in} = \int_{75}^{t_1} [17900 + 791(t-75)]\ \frac{F'(t)}{COP}\ dt$$

$$= \int_{75}^{t_1} [17900 + 791(t-75)]$$

$$\cdot\ [.235 + .0051(t-75)]F'(t)\ dt$$

Using the approximation to the annual distribution of mean temperatures given by (36), we compute these expressions at representative locations in each of the seven AIA weather zones in the United States, and compute the seasonal efficiency Q_{out}/Q_{in}. These values are then fitted empirically as a function of daily average heating and cooling degree-days. The empirical function is accurate to within 1 percent.

In the heating mode, the *COP* of heat pumps is approximately constant, with value 3.25, over the range where the unit is operational. Below an ambient mean temperature of 40°F, however, build-up of frost on the outdoor coil prevents operation, and backup heating is required. The usual system has electric resistance heating for extreme weather. We analyze this system. This method could be applied with obvious modifications to oil or gas backup units.

The energy output per hour for heating, averaged over the year, is:

$$Q_{out} = \int_{t_0}^{\tau} [800(\tau-t) - 1600]F'(t)\ dt$$

where τ is the thermostat setting and $800(\tau-t)-1600$ is the average heating load in Btuh over a day in the heating season (defined by $t < \tau$). If the input for days with $t < 40$ is provided by resistance heating, then:

$$Q_{in} = \int_{t_0}^{40} [800(\tau-t)-1600]F'(t) \, dt$$

$$+ \; \frac{1}{3.25} \int_{40}^{\tau} [800(\tau-t)-1600]F'(t) \, dt$$

$$= \; \frac{1}{3.25} \left(2.25 \int_{t_0}^{40} [800(\tau-t)-1600]F'(t) \, dt \right.$$

$$+ \; \int_{t_0}^{\tau} [800(\tau-t)-1600]F'(t) \, dt \;\;)$$

We compute Q_{out}/Q_{in} at representative locations in the seven AIA weather zones, using the approximation (36) to the temperature distribution. The resulting efficiencies are then approximated empirically as functions of heating and cooling degrees per day. The reported efficiencies are for a thermostat setting of 65 degrees. The empirical formula is accurate to within 4 percent.

CHAPTER 3

Estimation of a nested logit model for appliance holdings[1]

3.1. Introduction

In this chapter we estimate a discrete choice model for room air conditioning, central air conditioning, space heating, and water heating using data from the National Interim Energy Consumption Survey (NIECS) of 1978 and the Pacific Northwest (PNW) survey conducted in 1979-1980 by the Bonneville Power Administration. The reader is invited to consult Appendix A for references to the data sets and a detailed discussion of procedures used to prepare the data for econometric analysis. The use of microlevel disaggregated survey data to estimate discrete choice models of heating, ventilating, and air conditioning (HVAC) systems has been very recent, but one can find a few related models in Dubin and McFadden (1984), Brownstone (1980), Goett (1979), Hausman (1979), and McFadden, Kirschner, and Puig (1978). One of the virtues of the structure developed in this chapter is that it has been successfully embedded in a larger microsimulation system (the Residential End-Use Energy Policy System (REEPS)) for the purposes of policy forecasting (Goett (1979)).

Throughout this chapter, we follow an estimation framework that compares the results based on national-level data with those obtained using regional data. While the NIECS and PNW data sets are similar in content and scope (some 4000 households in each), important differences remain. During the early seventies, the Pacific Northwest region experienced average and marginal electricity prices that were very low by national average standards. Early projections of sustained growth in electricity demand necessitated increases in base-load generating capacity. However, the decision to provide additional capacity with nuclear plants has greatly increased the incremental cost of electricity.

[1] I acknowledge the research assistance of Paul Bjorn, who aided in the simulation experiments of Section 7. I further thank Steven Hensen for reading and providing comments on a preliminary draft.

It is plausible to assume that economic agents in a region with an inexpensive power source behave differently from consumers faced with viable economic trade-offs among alternative fuel sources. The comparison of results in the two data sets allows us to address the important issue of model transferability as well as lend support to our preferred specifications.

The chapter begins with a discussion (in Section 2) of the nested logit model of appliance choice and the particular tree extreme value structure used in our analysis. Ten alternative HVAC systems are considered and matched with actual operating and capital costs using an engineering thermal model that predicts heating and cooling loads. An important connection is thus established between the engineering and economic aspects of the choice problem.

Sections 3 and 4 validate the utility maximization hypothesis with the estimation of room air conditioning and water heating fuel choice conditional on the outcomes of HVAC system choice. We then develop (in Section 5) a nested logit model of space heating system choice. We consider the effect of income on the discount rate that annualizes capital cost, and explore the role of price expectation formation in the choice of HVAC systems. A time path of operating costs is matched to each household using historical state-level energy prices so that perfect, static, and adaptive expectations may be contrasted.

Section 6 estimates the full tree structure and discusses the determinants of central air conditioning choice, while Section 7 considers alternative conservation policies and alternative scenarios for the prices of electricity, natural gas, and fuel oil to forecast the path of durable good saturations from the present to the year 2000.

3.2. Nested logit model of appliance choice

This section describes the discrete choice model of alternative appliance portfolio combinations estimated from the National Interim Energy Consumption Survey and the Pacific Northwest Energy Survey. From the onset we desired to include as many of the major household appliances in the choice system as possible. We have concentrated on the choices of 19 alternative space heating and air conditioning systems, three water heating fuels, and the choice of room air conditioning. The possible combinations of appliance portfolios and the possible number of tree structures that might explain the observed choices are essentially limitless.

The empirical searches for nested logit forms that produce sensible results concentrated on a subset of the 19 alternative space heating systems. These alternatives form the trunk of the tree structure whose branches determine room air conditioning choice and the type of water heating fuel. The NIECS data reveal two important ingredients in this choice process: (1) the importance of eliminating gas heating system alternatives from the choice model when gas is not available as a fuel, and (2) the critical nature of scale effects, which manifest themselves in deleterious heteroscedasticity.

3.2.1. Natural gas availability

Whether natural gas is available obviously determines whether a household will install a gas heating system. If we include in the choice set an economically attractive gas alternative that is in fact unavailable, then we are sure to risk specification bias.

Unfortunately, measures of gas availability were not present within the NIECS data base. To construct a measure of gas availability, we followed two distinct procedures. First, we utilized a measure of gas availability that did exist for the Washington Center for Metropolitan Studies (WCMS) cross-sectional data. Given our ability to link locational information (at the level of primary sampling units) from one survey to the other, we were able to match the gas availability data from WCMS to NIECS. One problem is that gas availability is likely to be determined at the level of city blocks or in areas corresponding to secondary sampling units. This imparts a coarseness to a variable that is to be used at the individual level. A second difficulty with this procedure is that the survey year for WCMS was 1975 while the NIECS survey was conducted in 1978. This gap in time might affect our information about households making choices after 1975.

Our second procedure used related information about natural gas in two NIECS variables. The first variable indicates whether the household has gas appliances and is an index of their cumulative consumption. The second variable indicates whether the household uses natural gas for any purpose. We compute the percentage of households in each secondary sampling unit with either positive gas index or positive usage. Gas availability is accordingly assigned to each household in the relevant secondary sampling unit. The inherent weakness of this procedure is that it does not provide requisite historical information.

In early attempts to puzzle through the tree structure of appliance choice, we located a few cases in which a household would choose an oil heating system or an electric heating system when in fact a gas system would have been less expensive in terms of both operating and capital costs. For households in which we had previously assumed the availability of gas, this posed an interesting problem: Why do households choose dominated alternatives? The answer might be explained by variations in tastes, yet it was most often the case that gas was the dominating unchosen alternative and not other fuels. We resolved this issue by assuming that our discrete indicator of gas availability was incorrect for the household in question.

To improve our measure of gas availability, we made two modifications. The first change assumes that gas is available (irrespective of our previous assignment) if a particular household chooses gas. Our second modification works in quite the opposite direction: It imposes the condition that gas is not available whenever a household chooses an alternative that is dominated by a gas alternative.

The treatment of dominated alternatives to modify our assignment of gas availability may well introduce a degree of measurement error. Fortunately, the Pacific Northwest locational information does permit exact assignment of gas availability to each household at the point of dwelling construction.

In the estimation of a nested logit model of HVAC system choice, we regard the availability of gas as an essentially discrete phenomenon. We thus assume that when gas is available, gas HVAC systems are in the choice set. When gas is not available, the chosen alternative is presumed selected from alternatives that exclude gas systems. This approach differs from that of other researchers who introduce dummy interaction terms to indicate gas availability.[2]

3.2.2. Tree extreme value models

Figure 3.1 illustrates the nested logit choice model of four space heating systems with central air conditioning, six space heating systems without central air, water heating fuel choice, and room air conditioning.

[2]See Goett (1979) for an example of this approach.

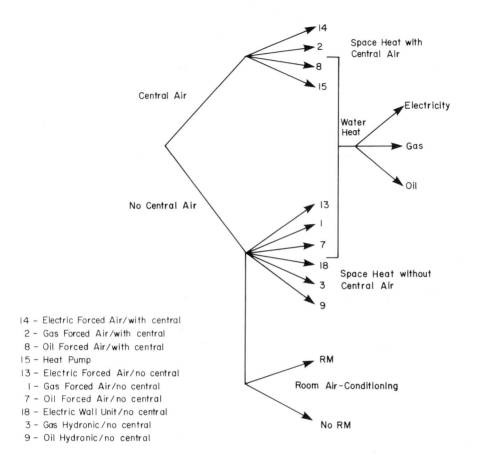

14 – Electric Forced Air/with central
2 – Gas Forced Air/with central
8 – Oil Forced Air/with central
15 – Heat Pump
13 – Electric Forced Air/no central
1 – Gas Forced Air/no central
7 – Oil Forced Air/no central
18 – Electric Wall Unit/no central
3 – Gas Hydronic/no central
9 – Oil Hydronic/no central

Figure 3.1

The postulated structure assumes that water heating choice is contingent on the choice of space heating system, that room air conditioning is chosen only when central air is not chosen, and that space heating choice is contingent on the choice of central air conditioning.

We arrive at this structure through a mixture of common sense and the accumulated wisdom of previous research. Unfortunately, a classical procedure to discriminate among specifications is not easily implemented given the non-nested nature of alternative tree structures. Standard errors reported in the NIECS estimation should be viewed with these considerations in mind. Estimation with the PNW data serves to test formally the structural hypothesis of interest and provides insight into the transferability of results from national to regional data. Furthermore, this approach lends support to the underlying utility maximization hypothesis. This hypothesis is likely to be violated in a region like the Pacific Northwest, where electricity is a cheap energy source and builders choose the heating system of least cost.

To derive a nested logit model for Figure 3.1, let Y_{wrsc} denote a positive measure of the desirability of alternatives indexed by $wrsc$ where w denotes water heating choice, r indicates room air conditioning choice, s indicates space heating choice, and c indicates central air choice. We specify a generating function $G[<Y_{wrsc}>]$ as the composition of four generating functions to reflect the levels of the tree in Figure 3.1:

$$G[<Y_{wrsc}>] = G^c[<G^s[<G^w[<G^r[<Y_{wrsc}>]>]>]>] \tag{1}$$

We take logistic generating forms for G^c, G^s, G^w, and G^r so that:

$$G^r[<Y_{rc}>] = \left[\sum_r Y_{rc}^{1/(1-\varphi)}\right]^{1-\varphi} \tag{2}$$

$$G^w[<Y_{wsc}>] = \left[\sum_w Y_{wsc}^{1/1-\sigma}\right]^{1-\sigma} \tag{3}$$

$$G^s[<Y_{sc}>] = \left[\sum_s Y_{sc}^{1/1-\delta_c}\right]^{1-\delta_c} \tag{4}$$

$$G^c[<Y_c>] = \sum_c Y_c \tag{5}$$

where φ, σ, and δ_c are unobserved scale factors. It follows that:

$$P_{wrsc} = (\partial \ln G^c / \partial \ln G^s) \cdot (\partial \ln G^s / \partial \ln G^w) \cdot (\partial \ln G^w / \partial \ln G^r) \cdot$$

$$(\partial \ln G^r / \partial \ln Y_{wrsc})$$

$$= P_c \cdot P_{s|c} \cdot P_{w|sc} \cdot P_{r|wsc} = (\partial \ln G / \partial \ln Y_{wrsc})$$

where P_{wrsc} denotes the probability of choosing portfolio combination $wrsc$, and $P_{j|k}$ denotes the conditional probability of choosing alternative j given that alternative k has been selected. To derive the structure in Figure 3.1, we assume that the probability of having room air conditioning conditional on HVAC choice is independent of heating system choice. Furthermore, we assume that the probability of water heating fuel choice is independent of room air conditioning choice. To impose this structure on the probability generating function G, we let $Y_{wrsc} = Y_{wsc} \cdot Y_{rc} \cdot Y_{sc} \cdot Y_c$. This model is consistent with the assumption that households maximize utility:

$$U_{wrsc} = V_{wrsc} + \varepsilon_{wrsc} \tag{6}$$

where $V_{wrsc} = \ln Y_{wrsc}$ denotes the strict utility of alternative $wrsc$ and $<\varepsilon_{wrsc}>$ have a joint generalized extreme value distribution. Note that the assumption:

$$Y_{wrsc} = Y_{wsc} \cdot Y_{rc} \cdot Y_{sc} \cdot Y_c$$

implies that strict utility may be written:

$$\ln Y_{wsc} + \ln Y_{rc} + \ln Y_{sc} + \ln Y_c = V_{wsc} + V_{rc} + V_{sc} + V_c$$

—a decomposition that exhibits the components of indirect utility. The generating function under the conditional independence assumption has the form:

$$G[Y_{wrsc}] = G^c[<Y_c G^s[<Y_{sc} G^w[<Y_{wsc}>]] \cdot G^r[<Y_{rc}>]>] \tag{7}$$

It is then possible to show that:

$$P_{r|c} = e^{V_{rc}/(1-\varphi)} / \sum_r e^{V_{rc}/(1-\varphi)} \equiv P_{r|wsc} \tag{8}$$

$$P_{w|sc} = e^{V_{wsc}/(1-\sigma)} / \sum_w e^{V_{wsc}/(1-\sigma)} \tag{9}$$

$$P_{s|c} = e^{(V_{sc} + J_{sc}(1-\sigma))/(1-\delta_c)} / \sum_s e^{(V_{sc} + J_{sc}(1-\sigma))/(1-\delta_c)} \tag{10}$$

$$P_c = e^{(J_c^s(1-\delta_c) + V_c + J_c^r(1-\delta))} / \sum_c e^{J_c^s(1-\delta_c) + V_c + J_c^r(1-\varphi))} \tag{11}$$

where:

$$J_{sc} \equiv \ln\left[\sum_w e^{V_{wsc}/1-\sigma} \right] \tag{12}$$

$$J_c^s \equiv \ln\left[\sum_s e^{(V_{sc} + J_{sc}(1-\sigma))/(1-\delta_c)} \right] \quad \text{and} \tag{13}$$

$$J_c^r \equiv \ln\left[\sum_r e^{V_{rc}/(1-\varphi)} \right] \tag{14}$$

The terms J_c^s, J_c^r, and J_{sc} are, respectively, the inclusive values of space heating choice given central air choice, room air choice given central air choice, and water heating choice given space heating and central air choice; the symbols $(1-\varphi)$, $(1-\delta_c)$, and $(1-\sigma)$ are the corresponding inclusive value coefficients.[3] Here we allow the inclusive value coefficient $(1-\delta_c)$ to be different depending on central air choice to reflect a possible dissimilarity in the degree of association in the space heating choice branches. Estimation of the central air conditioning choice model will identify the coefficients δ_c.

---

[3]The inclusive value coefficient is defined and interpreted in McFadden (1978).

3.3. Room air conditioning choice model

This section describes the estimation of the choice model for room air conditioning. The analysis considers room air conditioning only as an alternative to central air conditioning; it does not take into account either the choice of the number of room air conditioning units or their efficiencies. For details concerning these latter aspects of the choice process, see Brownstone (1980) and Hausman (1979).[4]

We begin with a review of the operating and capital costs that enter the utility maximization problem.

Our allocation of capital costs to central air conditioning units assumes that households purchase units of design capacity. Design capacity measures the thousands of Btu's per hour required to maintain a given household at summer design temperatures.[5] We follow the same procedure for room air conditioners and assume that room air conditioners are purchased to meet design cooling loads.

More precisely we assume that the total cooling load in the residence is distributed equally among the number of rooms in the residence; we then determine the capital costs (materials and installation) for providing one room air conditioning unit per room. Casual empiricism suggests this is a departure from average behavior, yet the assumption allows us to determine total capital costs in a manner that recognizes substantial returns to scale in purchasing larger air conditioning units. For additional details concerning the construction of room air conditioning costs, the reader is referred to Cowing, Dubin, and McFadden (1982).

Consistent with our determination of room air conditioning capital costs, we assume that the operating costs for room units distributing the total cooling load are identical to those for a central air conditioning system. This supposes (perhaps unrealistically) that room air conditioners operated in parallel are as efficient as central systems.

We would expect, other things being equal, that the probability of choosing room air conditioning given that the household does not have central air conditioning should increase with income and decrease as operating and capital costs increase. We have attempted an empirical

---

[4]The NIECS data provide information about the number of room air conditioners owned by the household and the number of rooms air-conditioned, but no information is available on individual room air conditioner efficiency. Estimation is confined to the NIECS data as air conditioning is not an important consideration in the Pacific Northwest.

[5]The thermal model of Chapter 2 provides direct estimates of air conditioning design capacity given household characteristics and location-specific temperature information.

specification in which these variables are interacted with the "purchase" alternative. In the "no purchase" alternative, we enter the number of household members and cooling degree-days, the latter a measure of the discomfort the household suffers in not having air conditioning. Table 3.1 presents the mean values of variables used in the room air conditioning choice model, while Table 3.2 presents the estimated models.

Table 3.1

Mean values for explanatory variables in room air
conditioning choice model (NIECS)

Variable	Description	Mean[a]
RMOPCST	Operating Cost for Room Air Conditioning (1967$)	49.22
RMCPCST	Capital Cost for Room Air Conditioning (1967$)	1231.
RMOPCST1	RMOPCST/(Base Load Usage)	0.00819
RMCPCST1	RMCPCST/(Base Load Usage)	3.33
CDD78	Cooling Degree-Days in 1978	1110
RINCOME	Income (1967$)/1000	10.38
NHSLDMEM	Number of Household Members	3.3

[a]Sample size 770 households corresponds to the set of single-family, detached, owner-occupied dwellings built since 1955 that do not have central air conditioning. A total of 591 of these homes appear in the nested logit model of HVAC system choice.

RINC1, CDD2, and PERS2 are RINCOME, CDD78, and NHSLDMEM interacted with alternative specific dummies for alternative one, alternative two, and alternative two respectively. A1 is the alternative-one specific dummy.

The operating and capital cost coefficients in Table 3.2 follow the pattern of results obtained by Goett (1979). Generally we observe that specifications that include operating and capital costs as well as cooling degree-days produce incorrect signs and insignificance in certain explanatory variables. It is possible to offer a few reasons for this result: 1) measurement error (which is likely given the assumptions made in assigning capital costs) would tend to bias the coefficient of capital cost towards zero, and 2) the desirability of room air conditioning is likely to be greatest when the cooling load is greatest, introducing a spurious correlation between operating costs, capital costs, and room air conditioning purchases.

To investigate the second effect in more detail, we present in Table 3.3 the room air conditioning choice model in which operating and capital costs are normalized by predicted base load usage (ACUEC). Note that the operating cost variable is now significant, but of wrong sign, while

Table 3.2

Binary logit model of room air conditioning choice
given no central air conditioning (NIECS)[a]

| Alternative 1 - Purchase Room Air Conditioning | | | 45.06 percent |
| Alternative 2 - Do Not Purchase Room Air Conditioning | | | 54.94 percent |

Variable Name	Logit Estimate	Standard Error	t-Statistic
RMOPCST	0.01139	0.004493	2.535
RMCPCST	-0.0001335	0.0002235	-0.5975
RINC1	0.03186	0.01478	2.156
CDD2	-0.0006152	0.0002111	-2.915
PERS2	0.02308	0.04907	0.4703
A1	-1.498	0.3393	-4.416

Auxiliary Statistics	At Convergence	At Zero
Log Likelihood	-467.9	-533.7
Percent Correctly Predicted[b]	68.18	50.00

[a]Estimation is by maximum likelihood using the QUAIL (Qualitative, Intermittent, and Limited Dependent Variable Statistical Program) developed by Daniel McFadden and Hugh Wills.

[b]A case is taken as being correctly predicted when the chosen alternative is forecast to have the highest probability of being chosen.

the normalized capital cost variable remains insignificant. The significance of the normalized operating cost variable may be attributable to a regional effect in which the largest average costs of room air conditioning are associated with regions in which there is a marginal electricity price peaking in summer. The summer peak rate is again associated with high average loads per customer due to the presence of very high ambient temperatures and a large percentage of homes using air conditioning.

Given the small change in log likelihood and percentage correctly predicted, we adopt the specification presented in Table 3.4 for use in the estimation of the HVAC choice tree. For the parameter estimates in Table 3.4, we construct the inclusive value of room air conditioning choice in the NIECS sample of 911 households. The mean value of RMINCV (room air conditioning inclusive value) is -.5041 with standard deviation .4022.

Table 3.3

Binary logit model of room air conditioning choice
given no central air conditioning normalized
operating and capital costs (NIECS)

Alternative 1 - Purchase
Alternative 2 - Do Not Purchase

Variable Name	Logit Estimate	Standard Error	t-Statistic
RMOPCST1	116.8	34.34	3.402
RMCPCST1	0.006826	0.004417	1.545
RINC1	0.03934	0.01439	2.734
CDD2	-0.001158	0.0001273	-9.098
PERS2	0.01186	0.04884	0.2429
A1	-2.813	0.4156	-6.768

Auxiliary Statistics	At Convergence	At Zero
Log Likelihood	-466.4	-533.7
Percent Correctly Predicted	68.70	50.00

Table 3.4

Binary logit model of room air conditioning
choice given no central air conditioning
no operating or capital costs (NIECS)

Alternative 1 - Purchase
Alternative 2 - Do Not Purchase

Variable Name	Logit Estimate	Standard Error	t-Statistic
RINC1	0.03765	0.01380	2.729
CDD2	-0.001104	0.0001190	-9.281
A1	-1.796	.2322	-7.732

Auxiliary Statistics	At Convergence	At Zero
Log Likelihood	-472.6	-533.7
Percent Correctly Predicted	70.26	50.00

3.4. Water heating choice model

This section describes the estimation of the choice model for water heating fuel using NIECS and PNW data. Related studies are Dubin and McFadden (1984) and Goett (1979). We begin with a review of the construction of operating and capital costs.

3.4.1. Water heating operating costs

We define the end-use service of water heating to be a gallon of heated water. To determine energy-to-service ratios (ESR), we used the March 1978 *Consumer Reports*, which reviewed 11 electric and 12 gas water heaters. *Consumer Reports* determined annual consumption in kwh per year and therms per year for electric and gas units respectively based on 100 gallons of hot water consumption per day. We used the mean value of annual consumption across models to calculate ESR by fuel type. For electric water heaters, the energy-to-service ratio is:

$$(10434.55 \ \frac{\text{kwh}}{\text{Yr.}}) \ (\frac{1 \ \text{Yr.}}{365 \ \text{days}}) \ (\frac{1 \ \text{day}}{100 \ \text{gal.}}) = 0.28588 \ \frac{\text{kwh}}{\text{gal.}}$$

and for gas water heaters, the energy-to-service ratio is:

$$(502.33 \ \frac{\text{therms}}{\text{gal.}}) \ (\frac{1 \ \text{Yr.}}{365 \ \text{days}}) \ (\frac{1 \ \text{day}}{100 \ \text{gal.}}) = 0.01376 \ \frac{\text{therms}}{\text{gal.}}$$

Following Dubin and McFadden (1984), we assume that oil water heaters are 74 percent as efficient as electric water heaters. Conversion to units of thousands of Btu's per gallon heated implies energy-to-service ratios: 1.376-gas, 0.97542-elec., and 1.318-oil. To determine average usage we use the relation:

Average annual usage in kwh for water heating

$$= (2819. + 360. \times (\text{NHSLDMEM}-2) + 360. \times$$

$$(\text{If NHSLDMEM equals 1})) + 365. \times 3.98 \times \text{HELDISHW}$$

This equation is discussed in Dubin and McFadden (1984). Note that NHSLDMEM and HELDISHW are, respectively, the number of household members and a dummy variable indicating that the household has a dishwasher. Finally, operating costs by fuel type are the product of (1) average annual usage, (2) the ratio of the ESR of the fuel under consideration to the ESR of the electric water heater, and (3) the price of the fuel at the point of house construction converted to real 1967 dollars.

3.4.2. Water heating capital costs

Construction of water heating capital costs requires a relationship between assumed capacity and structural characteristics of the dwelling and family. We follow the recommended practice ("Handbook of Buying 1978," *Consumer Research Magazine*) of relating capacity utilization to the number of bathrooms and the number of bedrooms (a proxy for number of persons). This relationship includes allowance for recovery rate differentials that occur between fuel types. Materials and installation costs for different-capacity water heaters are obtained from Means (1981). These estimates do not include the vent costs for each water heater. For vent costs, we consulted the *National Construction Estimator* (Solano Beach, Calif.: Craftsman Book Co., 1978) and determined that in 1981 dollars, material costs would be $18 while installation costs would be $26. The *National Construction Estimator* also indicated electrical contracting charges of $145 and $161 for water heaters with capacity less than and greater than 40 gallons respectively. These costs were added to the installation costs obtained from Means (1981). Finally, we have included all cost components that are conditional on the type of space heating system installed. When space heating type is gas or electric, the costs for materials and installation of an oil tank are included with the costs of oil water heating. When space heating type is gas or oil, an additional charge of $112 is added to the labor costs of the electric water heater due to the installation of increased amp service (*National Construction Estimator*, 1978). Other charges are assumed reflected in the cost of the heating systems.

3.4.3. Estimation of water heating choice model

Tables 3.5 and 3.6 present the mean values of NIECS and PNW variables used in the choice models as well as their descriptions. Estimation is based on a sample of households who live in single-family, owner-occupied dwellings built since 1955 and who choose either electric, gas, or oil water heaters.[6] As discussed above, the natural gas alternative is eliminated from the choice set whenever gas is unavailable to the household. Thus, in Table 3.5, the number of included observations drops

---

[6]The sample is additionally edited to eliminate infrequently selected space heating systems. This point will be taken up below.

Table 3.5

Mean values of variables in water heating
choice model (1967 dollars, NIECS)[a]

Variables	Alt[b]	Nobs	Description	Mean
WHOPCST	(1)	911	Water heating operating costs	113.40
WHOPCST	(2)	655	(by alternative)	23.69
WHOPCST	(3)	911		16.74
WHOPCST1	(1)	911	Water heating operating cost	0.02773
WHOPCST1	(2)	655	divided by usage	0.00582
WHOPCST1	(3)	911	(by alternative)	0.00406
WHCPCST	(1)	911	Water heating capital cost	201.50
WHCPCST	(2)	655	(by alternative)	130.90
WHCPCST	(3)	911		631.50
WHCPCST1	(1)	911	Water heating capital cost	0.05079
WHCPCST1	(2)	655	divided by usage	0.03336
WHCPCST1	(3)	911	(by alternative)	0.1621
SHE	(1)	911	(Space heating fuel: electricity) × (ALT1)	.2086
SHG	(2)	655	(Space heating fuel: gas) × (ALT2)	.8198

[a]Mean values for included alternatives.
[b]Electricity, natural gas, and fuel oil respectively.

from 911 in the electric and oil alternatives to 655 in the natural gas alternative. A similar effect is seen in Table 3.6.

We considered both binary and trinary specifications that used water heating operating and capital costs as well as space heating fuel-type dummies as explanatory variables. Models in which costs were not adjusted for scale provided generally wrong signs on variables and were difficult to interpret.

We present only specifications in which operating and capital costs are normalized by predicted utilization. Here normalized operating and capital costs are interpreted as the service price and capital cost per unit of service. Results for the NIECS and PNW data for binary choice models are given in Tables 3.7 and 3.8. Specifications that include electric, natural gas, and oil alternatives are given in Tables 3.9 and 3.10. All coefficients are highly significant and of the right sign. Generally we see that increases in operating and capital costs decrease the probability that an alternative is selected. The gas space heating system dummy in the second alternative is positive and very significant in all four models. Thus *the presence of a gas space heating system strongly influences the decision to choose gas as the fuel for water heating.*

Table 3.6

Mean values of variables in water heating
choice model (1967 dollars, PNW)[a]

Variables	Alt	Nobs	Description	Mean
WHOPCST	(1)	912	Water heating operating costs	58.94
WHOPCST	(2)	803	(by alternative)	36.49
WHOPCST	(3)	912		18.10
WHOPCST1	(1)	912	Water heating operating cost	0.01413
WHOPCST1	(2)	803	divided by usage	0.00861
WHOPCST1	(3)	912	(by alternative)	0.00423
WHCPCST	(1)	912	Water heating capital cost	213.10
WHCPCST	(2)	803	(by alternative)	135.50
WHCPCST	(3)	912		628.80
WHCPCST1	(1)	912	Water heating capital cost	0.05237
WHCPCST1	(2)	803	divided by usage	0.03302
WHCPCST1	(3)	912	(by alternative)	0.1555
SHE	(1)	912	(Space heating fuel: electricity) × (ALT1)	0.4287
SHG	(2)	803	(Space heating fuel: gas) × (ALT2)	0.4231

[a]Mean values for included alternatives.

The coefficient of the alternative specific dummy for the electric alternative is positive and significant across the specifications, indicating a preference for electric systems not accounted for by the other explanatory variables. The electric space heating system dummy, however, is not significant, which suggests the likely interaction of alternative specific and space heating system effects.

The ratio of capital to operating cost coefficients in the different specifications measures the real rate of transformation of capital cost into annualized life-cycle cost—in other words, the discount rate. The binary logit model including only electric and natural gas alternatives implies that these discount factors are 58.24 percent and 8.65 percent for the NIECS and PNW data respectively. The trinary models imply discount factors of 43.92 percent and 9.61 percent for the NIECS and PNW data respectively.

These differences in estimated discount rates are too large to be explained away through minor changes in the modeling assumptions. One likely explanation is that the historically low price of electricity in the Pacific Northwest leads to a high saturation of electric water heating systems, with much smaller attention paid to initial capital costs. This effect is further seen in the coefficients of capital costs in Tables 3.8 and 3.10. Although the qualitative results are very similar, it would not

Table 3.7

Binary logit model of water heating fuel choice given space
heating fuel choice—normalized costs (NIECS)

Alternative Label	Description	Frequency	Percent of Cases	Frequency Chosen	Percent Chosen
1	electric	640.0	100.0	118.0	18.44
2	natural gas	640.0	100.0	522.0	81.56

Variable Name	Logit Estimate	Standard Error	t-Statistic
WHOPCST1	-82.05	24.00	-3.419
WHCPCST1	-47.79	19.58	-2.441
A1	3.910	0.8756	4.465
SHE	-0.3276	0.5807	-0.5641
SHG	3.698	0.3839	9.632

Auxiliary Statistics	At Convergence	At Zero
Log Likelihood	-176.3	-443.6
Percent Correctly Predicted	91.25	50.00

appear that the national results are directly transferable to a region such as the Pacific Northwest, where energy prices have had such a profoundly different history.

We use the choice model in Table 3.9 in the estimation of the HVAC nested logit model. The calculation of inclusive values correctly accounts for the availability of natural gas. Thus, when gas is not available, the inclusive value corresponds to the electric and oil alternatives only.

3.5. Space heating system choice

Chapter 2 and Cowing, Dubin, and McFadden (1982) examine 19 alternative heating, ventilating, and air conditioning systems that provide combinations of heating and cooling capacity at design temperature conditions. We list the 19 alternative HVAC systems in Table 3.11. Seven of the 19 systems (numbers 4, 6, 10, 12, 16, 17, and 19) have very small sample frequencies and are not considered further in the NIECS data.

Additionally, we have been forced to eliminate gas and oil wall units from further study. These systems have both lower operating and capital costs than other HVAC systems. Wall units, however (especially gas and oil), are relatively infrequently selected. It is possible that aspects of these systems other than monetary considerations make them unattractive for

Table 3.8

Binary logit model of water heating fuel choice given space
heating fuel choice—normalized costs (PNW)

Variable Name	Logit Estimate	Standard Error	t-Statistic
WHOPCST1	-163.9	23.95	-6.844
WHCPCST1	-14.18	11.48	-1.236
A1	5.053	0.7133	7.083
SHE	0.7644	0.9538	0.8015
SHG	4.067	0.6279	6.477

Auxiliary Statistics	At Convergence	At Zero
Log Likelihood	-211.8	553.8
Percent Correctly Predicted	85.86	50.00

Table 3.9

Three-alternative multinomial logit model of water heating fuel choice
given space heating fuel choice—normalized costs (NIECS)[a]

Variable Name	Logit Estimate	Standard Error	t-Statistic
WHOPCST1	-104.1	17.41	-5.981
WHCPCST1	-45.72	8.535	-5.357
A1	2.043	0.5149	3.968
A2	-2.308	0.5983	-3.857
SHE	-0.2155	0.5248	-0.4107
SHG	3.722	0.3574	10.42

Auxiliary Statistics	At Convergence	At Zero
Log Likelihood	-272.6	-897.0
Percent Correctly Predicted	88.80	38.02

[a]Alternatives are electricity, natural gas, and fuel oil respectively.

installation, but it is more likely that the definitions of non-central systems used in the NIECS and PNW surveys are ambiguous.

Based on these considerations and various attempts with specifications that included these alternatives, we have opted to eliminate gas and oil wall units from the analysis. The remaining set of ten HVAC systems represent 911 single-family, owner-occupied detached dwellings built since 1955 with electric, gas, or oil water heating. Four of the ten alternatives include central air conditioning, and the sample is selected so that households choosing central air conditioning use electricity as the primary fuel (a small fraction of homes used gas central air condition-

Table 3.10

Three-alternative multinomial logit model of water heating fuel choice
given space heating fuel choice—normalized costs (PNW)

Variable Name	Logit Estimate	Standard Error	t-Statistic
WHOPCST1	-158.2	22.47	-7.039
WHCPCST1	-15.20	8.689	-1.750
A1	5.298	0.7048	7.517
A2	0.1803	1.001	0.1802
SHE	0.5535	0.7955	0.6958
SHG	4.147	0.5968	6.949

Auxiliary Statistics	At Convergence	At Zero
Log Likelihood	-239.7	-957.7
Percent Correctly Predicted	86.84	35.33

Table 3.11

Shares of alternative HVAC systems (NIECS)

HVAC System No.	Frequency[a]	Description	
1	0.2676	Gas Forced-Air	No Central Air
2	0.1234	Gas Forced-Air	Central Air
3	0.0639	Gas Hot Water	No Central Air
4	0.00496	Gas Hot Water	Central Air
5	0.1214	Gas Wall Unit	No Central Air
6	0.00396	Gas Wall Unit	Central Air
7	0.09118	Oil Forced-Air	No Central Air
8	0.02725	Oil Forced-Air	Central Air
9	0.06838	Oil Hot Water	No Central Air
10	0.00396	Oil Hot Water	Central Air
11	0.01933	Oil Wall Unit	No Central Air
12	0.00050	Oil Wall Unit	Central Air
13	0.01288	Electric Forced-Air	No Central Air
14	0.03023	Electric Forced-Air	Central Air
15	0.01685	Electric Heat Pump	Central Air
16	0.00149	Electric Hot Water	No Central Air
17	0.00000	Electric Hot Water	Central Air
18	0.05401	Electric Baseboard	No Central Air
19	0.00694	Electric Baseboard	Central Air

[a]Based on the sample of 2018 single-family, owner-occupied detached dwellings built since 1955.

ing). The two branches of the space heating choice model are illustrated in Figure 3.1 of Section 2.

Similar considerations in the Pacific Northwest data select ten alternative HVAC systems, which represent 912 single-family, owner-occupied detached dwellings built since 1955. Alternatives that include central air conditioning are excluded due to their small numbers. Table 3.12 presents the alternatives and their frequencies for the PNW data.

Table 3.12
Shares of alternative HVAC systems (PNW)

HVAC System No.	Frequency[a]	Description	
1	0.2549	Gas Forced-Air	No Central Air
3	0.0271	Gas Hot Water	No Central Air
5	0.0389	Gas Wall Unit	No Central Air
7	0.1647	Oil Forced-Air	No Central Air
9	0.0135	Oil Hot Water	No Central Air
11	0.0299	Oil Wall Unit	No Central Air
13	0.0761	Electric Forced-Air	No Central Air
15	0.0017	Electric Heat Pump	Central Air
16	0.0039	Electric Hot Water	No Central Air
18	0.2047	Electric Baseboard	No Central Air

[a]Based on the sample of 1773 single-family, owner-occupied detached dwellings built since 1955.

Tables 3.13 and 3.14 present the mean values of operating and capital costs by alternative and year of house construction in the NIECS data, while Tables 3.15 and 3.16 provide analogous means for the Pacific Northwest. All prices have been converted to 1967 dollars by cost indices using actual costs in the year built. An examination of Tables 3.13 and 3.15 indicates that in the post-1955 period, operating costs for oil systems were less expensive in real terms than operating costs for gas systems. This situation changed dramatically in the post-1972 period. Operating costs for electric systems are lower in the Pacific Northwest than corresponding average costs in the national data.

3.5.1. Water heating fuel and space heating system choice

Having selected a set of alternative HVAC systems, we now examine the cross-classification of water heating fuel and space heating system choices. These are presented in Tables 3.17 and 3.18 for the NIECS and PNW data respectively. A striking feature of these tables is the tendency for gas and oil water heating fuels to be selected most commonly with gas and oil space heating systems.

Table 3.13

Mean values of space heating operating costs by alternative
and year house built (1967 dollars, NIECS)

HVAC#	Alt	1955-1969	1960-1964	1965-1969	1970-1974	1975+	All Years
1	2	247	229	169	179	247	223
3	5	243	225	166	176	244	219
2	8	313	287	216	214	284	277
7	3	171	154	141	135	290	171
9	6	168	152	138	133	286	168
8	9	237	213	188	171	327	255
13	1	1142	897	672	631	743	911
15	10	718	553	395	389	471	566
14	7	1208	956	720	667	779	966
18	4	1056	825	615	578	685	840
Nobs		373	181	134	124	99	911

HVAC#	Description	
1	Gas Forced-Air	No Central Air
3	Gas Hot Water	No Central Air
2	Gas Forced-Air	Central Air
7	Oil Forced-Air	No Central Air
9	Oil Hot Water	No Central Air
8	Oil Forced-Air	Central Air
13	Electric Forced-Air	No Central Air
15	Electric Heat Pump	Central Air
14	Electric Forced-Air	Central Air
18	Electric Baseboard	No Central Air

Following Dubin and McFadden (1984), we tried a simple binary choice model of all electric versus all gas systems. In this model, operating and capital costs include the combined costs of forced-air and water heating. Table 3.19 presents the estimated choice models in normalized forms for the NIECS data, while Table 3.20 presents corresponding results for the Pacific Northwest. We note again that prenormalization of operating and capital costs is a necessary step to achieve sensible results. For direct comparison to the Dubin and McFadden (1984) model, we include an additional explanatory variable that interacts capital cost with real income. This variable permits a discount factor specification that varies linearly in income. The NIECS discount factor (using the coefficient estimates in the normalized model) is (19.79 - 0.81 (RINCOME/1000)) while the PNW discount factor is (61.61 - 5.14 (RINCOME/1000)). Dubin and McFadden (1984), using national data in the Washington Center for Metropolitan Studies Energy Survey, estimate the linear-in-income discount factor to be (37.93 - 1.028 RINCOME).

Table 3.14

Mean values of space heating capital costs by alternative
and year house built (1967 dollars, NIECS)

HVAC#	Alt	1955-1959	1960-1964	1965-1969	1970-1974	1975+	All Years
1	2	1110	1055	1017	1063	1043	1072
3	5	2279	2327	2343	2623	2594	2379
2	8	2057	1880	1902	1839	1786	1940
7	3	1843	1698	1609	1637	1595	1725
9	6	2818	2809	2795	3076	3027	2871
8	9	2489	2261	2256	2187	2123	2329
13	1	918	876	843	880	862	887
15	10	4935	4514	3920	4353	4504	4576
14	7	1938	1879	1824	1863	1815	1886
18	4	912	889	837	917	929	899
Nobs		373	181	134	124	99	911

While this estimated relationship is bracketed by the NIECS and PNW results, it is not surprising that it is closer to the NIECS estimates. The evidence in the cross-classification tables and the assumptions under which costs are assigned would tend, however, to reject these models in favor of a richer specification in which each HVAC system is expressed as an individual alternative and in which water heating choice is estimated conditionally upon space heating system choice.

3.5.2. Nested logit model of space heating system choice

The variables SHOPCST and SHCPCST represent the operating and capital costs of ten alternative HVAC systems. These variables are calculated using annual predictions of usage and capacity developed in the thermal model. Operating and capital costs for alternatives that include air conditioning reflect additional costs associated with the central air conditioner and any economies that result from shared costs.[7] The variables SHOPCST1 and SHOPCST2 are SHOPCST divided by two scaling factors: predicted usage (SHUECE) and the operating cost of an electric baseboard heating system respectively. The empirical analysis indicates that either method of scaling provides adequate results. Furthermore, the scaled variables have strong intuitive appeal. To see this, let us consider the operating cost of system j:

[7]For details the reader is referred to Cowing, Dubin, and McFadden (1982).

Table 3.15

Mean values of space heating operating costs by alternative
and year house built (1967 dollars, PNW)

HVAC#	Alt	1955-1959	1960-1964	1965-1969	1970-1974	1975+	All Years
1	2	536	329	288	218	250	352
3	5	527	324	284	215	246	346
5	8	493	303	265	200	229	323
7	3	205	214	200	171	278	217
9	6	202	211	197	168	273	213
11	9	189	197	184	157	254	199
13	1	799	729	633	477	361	614
15	10	311	256	245	173	202	232
16	7	786	718	624	470	354	604
18	4	738	673	584	438	330	566
Nobs		282	136	150	140	204	912

HVAC#	Description	
1	Gas Forced-Air	No Central Air
3	Gas Hot Water	No Central Air
5	Gas Wall Unit	No Central Air
7	Oil Forced-Air	No Central Air
9	Oil Hot Water	No Central Air
11	Oil Wall Unit	No Central Air
13	Electric Forced-Air	No Central Air
15	Electric Heat Pump	Central Air
16	Electric Hot Water	No Central Air
18	Electric Baseboard	No Central Air

$$SHOPCST_j = (SHUECE)(D_j)(1/COP_j) \cdot P_j \quad \text{where:}$$

$SHOPCST_j$ = operating cost of system j

$SHUECE$ = base load usage (delivered Btu's)

D_j = delivery system adjustment factor

COP_j = coefficient of performance for system j

P_j = price of fuel used by system j

Note that the electric baseboard system (HVAC no. 18) has a coefficient of performance equal to one, that it has a delivery factor one, and that because it uses electricity its operating cost is ($SHUECE \cdot P_e$). The normalization rules imply that:

$$SHOPCST1_j = (D_j)(1/COP_j)P_j$$

$$SHOPCST2_j = (D_j)(1/COP_j)(P_j/P_e)$$

Table 3.16
Mean value of space heating capital costs by alternative
and year house built (1967 dollars, PNW)

HVAC#	Alt	1955-1959	1960-1964	1965-1969	1970-1974	1975+	All Years
1	2	1129	1099	1065	1106	999	1081
3	5	2417	2562	2599	2851	2466	2546
5	8	1063	1160	1185	1322	1173	1162
7	3	1870	1750	1656	1687	1565	1721
9	6	2964	3052	3054	3312	2913	3034
11	9	1467	1520	1517	1651	1489	1516
13	1	951	933	890	927	828	907
15	10	5060	4854	4697	4655	3738	4612
16	7	3041	3072	3042	3247	2844	3033
18	4	978	992	969	991	825	946
Nobs		282	136	150	140	204	912

Table 3.17
Cross-classification of water heating fuel and
space heating system choice (NIECS)

			Water Heating Fuel		
HVAC#	Space Heating System	Description	Electric	Gas	Oil
1	Electric Forced-Air	No Central Air	21	0	0
2	Gas Forced-Air	No Central Air	23	271	0
3	Oil Forced-Air	No Central Air	79	9	11
4	Electric Baseboard	No Central Air	75	3	0
5	Gas Hot Water	No Central Air	1	56	0
6	Oil Hot Water	No Central Air	6	3	33
7	Electric Forced-Air	Central Air	55	5	0
8	Gas Forced-Air	Central Air	12	174	0
9	Oil Forced-Air	Central Air	39	1	3
10	Electric Heat Pump	Central Air	31	0	0

(Based on a sample of 911 households.)

The first normalization method replaces operating cost by an efficiency adjusted price, while the second method further scales all costs by the price of electricity. The efficiency adjusted price SHOPCST1$_j$ is related to the price of comfort because the latter is SHOPCST1$_j$ multiplied by the marginal increase in usage required to change the thermostat setting 1 degree. For a given household, this quantity would be constant across alternatives and would change all normalized operating costs in a proportional manner. Empirical results obtained using the calculated price of comfort rather than normalized operating costs were very similar. The normalized variables have the additional advantage of being sensible

Table 3.18
Cross-classification of water heating fuel and
space heating system choice (PNW)

HVAC#	Space Heating System	Description	Water Heating Fuel		
			Electric	Gas	Oil
1	Electric Forced-Air	No Central Air	118	0	0
2	Gas Forced-Air	No Central Air	104	198	0
3	Oil Forced-Air	No Central Air	136	3	2
4	Electric Baseboard	No Central Air	262	1	1
5	Gas Hot Water	No Central Air	5	26	0
6	Oil Hot Water	No Central Air	13	0	1
7	Electric Hot Water	No Central Air	5	1	0
8	Gas Wall Unit	No Central Air	10	4	0
9	Oil Wall Unit	No Central Air	18	0	1
10	Electric Heat Pump	Central Air	3	0	0

(Based on a sample of 912 households.)

on econometric grounds, because the unobserved component of utility would otherwise be heteroscedastic. Furthermore, the normalization has psychometric appeal given the assumption that households evaluate relative rather than absolute system costs.

Table 3.21 presents the estimation of space heating choice models based on subsets of the ten alternative systems. Specifications 3 and 4 present subsets of the alternatives appearing in specifications 1 and 2. Similarly, specifications 5 and 6 are nested cases of specifications 7 and 8. Departures from the assumption of independence of irrelevant alternatives (IIA) or from the *a priori* grouping of alternatives should be detectable in significant changes in the estimated coefficients. We maintain that grouping space heating systems into subsets with and without central air conditioning is sensible given the distinct nature of unobserved effects that characterize each technology.

The results of the estimation are quite sensible in terms of significance and sign. Further, there do not appear to be any obvious departures from our selected groupings of alternatives. Without extensive specification testing, it would be difficult to reject the assumption of IIA or evaluate its consequences for point estimation.[8]

We find the inclusive value coefficient to be insignificant across the various specifications. This is not inconsistent with the assumption of

[8]McFadden, Tye, and Train (1978) and Hausman and McFadden (1984) discuss two tests of the I.I.A. property.

Table 3.19

Binary logit model of space heating and water heating fuel choice—
normalized costs (NIECS)

Description	Alternative Label	Frequency	Percent of Cases	Frequency Chosen	Percent Chosen
Electric Forced-Air and Water	1	683.0	100.0	182.0	26.65
Gas Forced-Air and Water	2	683.0	100.0	501.0	73.35

Means of Independent Variables

Alt. Label	OPCST1	CPCST1	CAPINC1	A1
1	1.0000	1.334	15.33	1.000
2	0.2604	1.517	17.39	0.000

Variable Name	Logit Estimate	Standard Error	t-Statistic
OPCST1	-12.15	1.187	-10.24
CPCST1	-2.704	1.110	-2.435
CAPINC1	0.09854	0.09098	1.083
A1	7.558	0.9421	8.023

Auxiliary Statistics	At Convergence	At Zero
Log Likelihood	-267.8	-473.4
Percent Correctly Predicted	85.80	50.00

random utility maximization. It indicates that consumers respond to the maximum utility of possible water heating fuel alternatives in their selection of a space heating system.

Given the small differential in mean water heating inclusive value across space heating fuel types, it is likely that there is significant interaction between the inclusive value variable and the alternative specific dummies. This is further confirmed by the fact that the estimated coefficients of operating and capital costs remain robust even when the inclusive value coefficient is constrained to be zero (models 13 and 14 of Table 3.21).

To explore this interaction hypothesis, we have estimated specifications 9, 10, 11, and 12 in Table 3.21. These models eliminate the alternative specific effect for oil alternatives. The estimates of the inclusive value coefficients in specifications 9 and 10 remain insignificant. The hypothesis that the estimated inclusive value coefficients in the central air conditioning branch (specifications 11 and 12) equal one cannot,

Table 3.20
Binary logit model of space heating and water
heating fuel choice—normalized costs (PNW)

Description	Alternative Label	Frequency	Percent of Cases	Frequency Chosen	Percent Chosen
Electric Forced-Air and Water	1	527.0	100.0	332.0	63.00
Gas Forced-Air and Water	2	527.0	100.0	195.0	37.00

Means of Independent Variables

Alt. Label	OPCST1	CPCST1	CAPINC1	A1
1	1.0000	2.242	22.62	1.000
2	0.6260	2.500	25.23	0.000

Variable Name	Logit Estimate	Standard Error	t-Statistic
OPCST1	-3.996	0.4801	-8.322
CPCST1	-2.462	1.032	-2.385
CAPINC1	0.1265	0.06557	1.929
A1	2.016	0.3446	5.850

Auxiliary Statistics	At Convergence	At Zero
Log Likelihood	-270.3	-365.3
Percent Correctly Predicted	75.71	50.00

however, be rejected under either normalization procedure. There is no *a priori* reason to expect that the inclusive value coefficients should differ in the two branches. Any difference in the two estimates of the inclusive value coefficient could be explicable only by differences in the degree of intracorrelations in each space heating choice cluster. The sequential estimation procedure cannot impose the constraint that the inclusive value coefficients be equal. It is thus uncertain whether water heating choice given space heating choice is the indicated specification. We therefore adopt the strategy of excluding the water heating choice inclusive value in the space heating choice estimation. The argument is that the differences in the inclusive values are small and are adequately captured in the alternative specific effects.

Estimation of discount factors appears robust across specifications. The discount factors are much lower for the set of alternatives that includes air conditioning as compared with the set of alternatives that does not include air conditioning. This may be a reflection of shared cost

Table 3.21
Estimation of space heating choice model (NIECS)[a]

Model	Alternative Label	Frequency	Percent of Cases	Frequency Chosen	Percent Chosen
Specifications 1, 2, 9, 10, and 13	1	591.0	100.0	21.00	3.55
	2	424.0	71.7	294.00	69.34
	3	591.0	100.0	99.00	16.75
	4	591.0	100.0	78.00	13.20
	5	424.0	71.7	57.00	13.44
	6	591.0	100.0	42.00	7.11
Specifications 3, 4	1	414.0	100.0	21.00	5.07
	2	334.0	80.7	294.00	88.02
	3	414.0	100.0	99.00	23.91
Specifications 5, 6	7	289.0	100.0	60.00	20.76
	8	223.0	77.2	186.00	83.41
	9	289.0	100.0	43.00	14.88
Specifications 7, 8, 11, 12, and 14	7	320.0	100.0	60.00	18.75
	8	231.0	72.2	186.00	80.52
	9	320.0	100.0	43.00	13.44
	10	320.0	100.0	31.00	9.69

[a]Total cases 911.

components in all-electric HVAC systems. It should be further noted that these estimates are considerably lower than those obtained in non-nested or binary specifications (Hausman (1979), Dubin and McFadden (1984)).

To explore the validity of these conjectures, we reestimate the space heating choice model using the Pacific Northwest data. Here we confine ourselves to the first six NIECS alternatives that do not include central air conditioning. Table 3.22 presents five alternative specifications. Specifications 1 and 2 correspond to specifications 1 and 2 of Table 3.21 while specifications 3 and 4 correspond to specifications 9 and 10 of Table 3.21. Finally, we estimate a non-normalized model in specification 5.

Qualitatively, the results in the PNW data are similar to those obtained in the NIECS data. Remarkably, the second normalization is now identified to be superior to first normalization. Under the second normalization (specifications 2 and 4), we find larger likelihoods at convergence and significant water heating inclusive value coefficients that lie in the unit interval.

Estimated discount factors are somewhat larger in the PNW data than those in the NIECS data. This is possibly a reflection that difficulties in

Table 3.21—continued

Model: Alt:	1 123456	2 123456	3 123	4 123	5 789	6 789	7 789 10
SHOPCST1	-722.9 (79.94)	- -	-901.1 (141.7)	- -	-439.2 (104.9)	- -	-324.6 (88.03)
SHCPCST1	-46.93 (20.85)	- -	-71.91 (39.09)	- -	-26.80 (17.90)	- -	-9.519 (8.527)
SHOPCST2	- -	-6.108 (1.158)	- -	-8.105 (1.625)	- -	-7.907 (1.757)	- -
SHCPCST2	- -	-0.6385 (0.1337)	- -	-.7874 (.2704)	- -	-.1730 (.1150)	- -
WHINCV	-0.2654 (0.3151)	0.1867 (0.3204)	0.1510 (0.4157)	0.1978 (0.4663)	0.3262 (0.5036)	0.5105 (0.4833)	0.9263 (0.4650)
A1	2.627 (.6676)	2.578 (1.167)	1.929 (0.6897)	3.270 (1.420)	- -	- -	- -
A2	3.175 (.8070)	1.589 (0.8344)	1.602 (1.002)	1.347 (1.167)	- -	- -	- -
A3	.3949 (.2799)	.04288 (.2460)	- -	- -	- -	- -	- -
A4	3.556 (.6294)	3.206 (1.064)	- -	- -	- -	- -	- -
A5	2.097 (.7646)	0.8637 (.8030)	- -	- -	- -	- -	- -
A6	- -	- -	- -	- -	- -	- -	- -
A7	- -	- -	- -	- -	2.911 (0.7229)	6.736 (1.520)	1.495 (0.4368)
A8	- -	- -	- -	- -	1.864 (1.074)	1.704 (1.053)	.00634 (.8184)
A9	- -	- -	- -	- -	- -	- -	-0.6996 (0.4429)
Log Likelihood	-570.9	-582.2	-136.1	-137.9	-148.3	-143.7	-232.7
Percent Correctly Predicted	64.47	64.97	89.37	88.89	79.24	79.93	70.94
Discount Factor	6.49	10.45	7.98	9.71	6.10	2.19	2.93

WHINCV is the inclusive value estimated in the water heat choice model. The variables SHCPCST1 and SHCPCST2 are SHCPCST divided by the scaling factors used in the normalization of operating costs. (Standard errors in parenthesis.)

Table 3.21—continued

Model: Alt	8 789 10	9 123456	10 123456	11 789 10	12 789 10	13 123456	14 789 10
SHOPCST1	- -	-728.0 (79.86)	- -	-233.4 65.40	- -	- -	- -
SHCPCST1	- -	-70.06 (14.03)	- -	-1.603 (5.950)	- -	- -	- -
SHOPCST2	-3.110 (0.8889)	- -	-6.066 (1.130)	- -	-2.210 .6415	-6.420 (1.031)	-4.499 (.8240)
SHCPCST2	-0.0578 (0.0585)	- -	-0.6538 (0.1014)	- -	-.000234 (.0412)	-0.6400 (0.1336)	-0.08259 (0.05766)
WHINCV	1.414 (.4168)	-0.2971 (0.3148)	0.1883 (0.3203)	1.277 (0.4161)	1.621 (0.3953)	- -	- -
A1	- -	2.100 (0.5551)	2.485 (1.036)	- -	- -	2.868 (1.058)	- -
A2	- -	2.742 (0.7468)	1.534 (0.7740)	- -	- -	2.030 (.3564)	- -
A3	- -	- -	- -	- -	- -	.04673 (.2452)	- -
A4	- -	3.004 (0.4973)	3.114 (0.9228)	- -	- -	3.468 (.9660)	- -
A5	- -	1.949 (0.7574)	0.8305 (0.7806)	- -	- -	1.308 (.2586)	- -
A6	- -	- -	- -	- -	- -	- -	- -
A7	2.045 (0.5552)	- -	- -	1.560 (0.4224)	1.959 (0.5428)	- -	2.694 (0.5376)
A8	-0.9816 (.7978)	- -	- -	-0.0868 (.8185)	-0.7589 (.7796)	- -	1.296 (0.4358)
A9	-0.7027 (0.4682)	- -	- -	- -	- -	- -	-1.309 (0.4387)
Log Likelihood	-233.4	-571.9	-582.2	-234.0	-234.5	-582.4	-239.8
Percent Correctly Predicted	70.94	65.31	65.14	71.25	71.25	65.14	69.69
Discount Factor	1.86	9.62	10.78	0.69	0.01	9.97	1.84

(Standard errors in parenthesis.)

Table 3.22
Estimation of space heating choice model (PNW)[a]

	Alternative Label	Frequency	Percent of Cases	Frequency Chosen	Percent Chosen
Specifications	1	752.0	100.0	118.0	15.69
1, 2, 3, 4, 5,	2	574.0	76.3	194.0	33.80
6, and 7	3	752.0	100.0	141.0	18.75
	4	752.0	100.0	264.0	35.11
	5	574.0	76.3	21.0	3.66
	6	752.0	100.0	14.0	1.86

[a]Total cases 912.

the measurement of gas availability in the NIECS data have been adequately corrected in the PNW data. Specification 5 of Table 3.22 strengthens our conclusions regarding the importance of prenormalization of operating and capital costs.

To summarize our findings in the comparison of the NIECS and PNW estimated choice models, we note that while qualitatively they reinforce one another, there are strong regional effects that make it impossible to transfer parameter estimates from one data set to the other. The national data reveal small discount factors in the space heating choice models, while much larger discount factors prevail in the water heating choice models. This pattern is reversed in the Pacific Northwest, where historically low energy costs are not as heavily weighted in the utility maximization.

3.5.3. Space heating system choice—income effects

We now investigate the significance of income on the choice of space heating systems using the Pacific Northwest data. Following the specification of Dubin and McFadden (1984), we do this by creating an interaction variable that is the product of normalized capital cost and real income. The estimated choice models are presented in Table 3.23.

The coefficient estimates of operating and capital cost are interpreted as linear-in-income discount factors. Each relationship is precisely determined and provides robust estimates of the income effect. A rule of thumb would indicate that each increase of $1000 in real income decreases the discount factor by 1 percent. That lower income levels are associated with higher discount rates is consistent with the hypothesis

Table 3.22—continued
Specifications that include water
heating inclusive value (PNW)

Model: Alt:	1 123456	2 123456	3 123456	4 123456
Variable				
SHOPCST1	-576.9 (67.23)	- -	-582.7 (68.49)	- -
SHCPCST1	-87.09 (22.77)	- -	-164.01 (17.27)	- -
SHOPCST2	- -	-1.8212 (0.3523)	- -	-1.569 (0.3471)
SHCPCST2	- -	-0.4968 (0.07967)	- -	-0.6902 (0.06792)
WHINCV	0.2165 (1.148)	0.7203 (0.1778)	0.1890 (0.1900)	0.8007 (0.1787)
A1	2.082 (0.5343)	0.6862 (0.5604)	0.2152 (0.3441)	-.9615 (.3821)
A2	2.035 (0.4763)	0.9164 (0.4579)	0.2731 (0.2726)	-.5391 (.2726)
A3	1.521 (.3484)	1.214 (0.3225)	- -	- -
A4	2.659 (0.5296)	1.230 (0.5450)	0.7402 (0.3262)	-.4330 (.3564)
A5	0.7478 (0.4238)	-0.1123 (0.4426)	0.1891 (0.3448)	-1.173 (.3427)
Log Likelihood	-951.5	-945.2	-27.24	-38.99
Percent Correctly Predicted	44.02	44.68	64.86	52.50
Discount Factor	15.10	27.28	28.16	43.99

(Standard errors in parenthesis.)

that the capital market constrains low income individuals to purchase high operating and low capital cost systems.

3.5.4. *The role of price expectation formation in the choice of heating systems*

In each of the specifications considered above, we have maintained the hypothesis that expectations of fuel prices were static. We have implicitly

Table 3.22—continued
Specifications that do not include
water heating inclusive value (PNW)

Model:	5	6	7
Alt:	123456	123456	123456
Variable			
SHOPCST	-	-	-.004832
	-	-	(.0003588)
SHCPCST	-	-	.001072
	-	-	(.0002014)
SHOPCST1	-	-699.3	-
	-	(54.98)	-
SHCPCST1	-	-99.11	-
	-	(22.71)	-
SHOPCST2	-3.082	-	-
	(0.2875)	-	-
SHCPCST2	-0.4301	-	-
	(0.07369)	-	-
A1	2.500	2.500	6.587
	(.4480)	(.4733)	(.6131)
A2	2.540	2.595	6.491
	(.3668)	(.4249)	(.5678)
A3	1.437	1.432	4.087
	(.3123)	(.3456)	(.4557)
A4	2.947	3.035	7.226
	(.4380)	(.4742)	(.6091)
A5	1.391	1.386	2.418
	(.3364)	(.3396)	(.3628)
Log Likelihood	-1107.	-1094.	-1078.
Percent Correctly Predicted	44.94	44.37	47.47
Discount Factor	13.96	14.17	-22.19

(Standard errors in parenthesis.)

assumed that the real price of alternative fuels would remain constant at the levels that prevailed at the point of dwelling construction. It is plausible, however, to assume that consumers view a trade-off between initial capital cost and the life-cycle cost of durable service. The components of life-cycle cost are determined primarily through the price of the indirectly demanded fuel input. Because these prices are unknown to

Table 3.23
Estimation of space heating choice model specifications
with income interactions (PNW)[a]

Alternative Label	Frequency	Percent of Cases	Frequency Chosen	Percent Chosen
1	733.0	100.0	102.0	13.92
2	646.0	88.1	254.0	39.32
3	733.0	100.0	113.0	15.42
4	733.0	100.0	226.0	30.83
5	646.0	88.1	26.0	4.03
6	733.0	100.0	12.0	1.64

[a]Total cases 912. Some observations are missing due to incomplete data on real income.

the consumers at the point of dwelling construction, one must assume that some expectation formation mechanism determines future prices.

Rather than postulate a particular expectation formation pattern, we specify alternative choice models that include past, present, and future operating costs. We then test the hypothesis that only past and present operating costs are jointly significant and interpret the estimated coefficients as an "adaptive expectation system." A test of the hypothesis that current operating costs are solely significant reveals a "static expectation system." Finally, the joint significance of future operating costs might be interpreted as "perfect foresight."

Each household in the NIECS data is located at the level of its primary sampling unit. Primary sampling units are matched to the State Energy Data Base (SEDS) from which we access historical energy prices. The NIECS data base provides categorical information on the year of dwelling construction. The interval changes from five-year periods in the years 1960 to 1975 to ten-year periods prior to 1960. It is not possible to determine precisely the year of dwelling construction prior to 1974 and it is therefore impossible to define exact lag and lead lengths.

We adopt the strategy of using five-year periods to define leads and lags. During each period a representative year is selected. We make the assumption that the real price of energy remains constant in each year-built category. Table 3.24 summarizes the assignment of selected years to categories and defines the lead and lag structure. We see from Table 3.24 that our assignments are further disturbed for houses built during the seventies. The difficulty arises because the SEDS data base extends from 1928 to 1980 and we cannot easily match additional information to define precise two-period leads.

Table 3.23—continued

Model:	1	2
Alt:	123456	123456
Variable		
SHOPCST1	-696.3	-
	(58.99)	-
SHCPCST1	-152.1	-
	(29.29)	-
SHOPCST2	-	-3.097
	-	(0.3100)
SHCPCST2	-	-0.7028
	-	(0.1109)
SHCAP1[b]	7.997	-
	(1.915)	-
SHCAP2[c]	-	0.03086
	-	(0.0066)
A1	3.144	2.845
	(0.5368)	(.4903)
A2	3.168	2.836
	(0.4827)	(.4013)
A3	1.814	1.633
	(0.3915)	(.3439)
A4	3.666	3.266
	(0.5371)	(.4791)
A5	1.525	1.453
	(0.3715)	(.3650)
Log Likelihood	-911.9	-924.4
Percent Correctly Predicted	45.98	45.98

[b]SHCAP1 = RINCOME * SHCPCST1.
[c]SHCAP2 = RINCOME * SHCPCST2.
(Standard errors in parenthesis.)

Specification	Discount Factors (Percent)
1	21.84 - 1.1485 × (RINCOME/1000)
2	22.69 - 0.9664 × (RINCOME/1000)

The estimation of the space heating choice model with past, present, and future operating costs is given in Table 3.25. We consider two specifications: specification 1 includes both leads and lags while specification 2 constrains lead price coefficients to be zero.[9]

[9]Normalization method 2 is used in the estimation of specifications 1 and 2.

Table 3.24

Definition of past and future operating costs
structure of leads and lags (NIECS)

Lags		Year built	Leads	
8	9	0	1	2
67	72	77	77	77
62	67	72	77	77
57	62	67	72	77
52	57	62	67	72
47	52	57	62	67

Key: 47: 1945-1949; 52: 1950-1954; 57: 1955-1959; 62: 1960-1964; 67: 1965-1969; 72: 1970-1974; 77: 1975-1979.

A comparison of the two models indicates the significance of future prices. On the other hand, past prices do not appear jointly significant. On balance it would appear that there is evidence to support the hypothesis that future prices matter to consumers. The model with future prices has greater predictive power than previous specifications that impose the static expectation hypothesis.

A pattern emerges in which an increase in next-period price will decrease the probability of selection for alternatives of that fuel type, while an expected increase in two-period forward prices works in the opposite direction. If this structure receives further empirical support, it should have important implications for both policy and prediction.

3.6. Central air conditioning choice

This section presents the estimation of the central air conditioning choice model. From equation (11) of Section 2, we see that the probability of air conditioning choice depends on the inclusive value of room air conditioning (when central air is not chosen), the inclusive values of space heating choice given air conditioning choice, and on other attributes of the utility of purchasing an air conditioning system. We follow the formulation of Chapter 1 and use income and cooling degree-days interacted with the first and second alternatives (central vs. noncentral) as determinants of the utility associated with either alternative. The inclusive value of room air conditioning choice interacts with the second alternative, as does the inclusive value of space heating choice given no central air conditioning. The inclusive value of space heating choice

Table 3.25
Estimation of space heating choice model with past and
future operating cost measures (NIECS)

	Alternative Label	Frequency	Percent of Cases	Frequency Chosen	Percent Chosen
Specifications 1 and 2	1	580.0	100.0	21.0	3.62
	2	396.0	68.3	288.0	72.73
	3	580.0	100.0	99.0	17.07
	4	580.0	100.0	78.0	13.45
	5	396.0	68.3	52.0	13.13
	6	580.0	100.0	42.0	7.24

Model:	1	2
Alt:	123456	123456

Variable		
SHCST82	-1.483 (0.9428)	-1.585 (0.9426)
SHCST92	2.714 (2.291)	2.353 (2.341)
SHCST02	-5.734 (2.541)	-6.012 2.154)
SHCST12	-3.687 (1.886)	- -
SHCST22	3.633 (1.481)	- -
A1	1.728 (1.187)	2.173 (1.155)
A2	2.159 (.3618)	2.110 (.3631)
A3	0.033 (.248)	0.056 (.247)
A4	2.483 (1.079)	2.864 (1.054)
A5	1.401 (0.2684)	1.334 (0.2654)
SHCPCST1	-0.6511 (0.1368)	-0.6356 (0.1373)
Log Likelihood	-550.6	-553.5
Percent Correctly Predicted	66.21	66.21

(Standard errors in parenthesis.)

given central air conditioning interacts with the first alternative. The results of the estimation are presented in Table 3.26.

Table 3.26
Binary logit model of central air conditioning
choice with inclusive value terms (NIECS)

	Alternative Label	Frequency of Cases	Percent Chosen	Frequency Chosen	Percent
Central AC	1	911.0	100.0	320.0	35.13
No Central AC	2	911.0	100.0	591.0	64.87

Means of Independent Variables:

Alt. Label	SHINCVC	SHINCVNC	RMINCV	RINCOME1	CDD2
1	-0.7905	0.0	0.0	12.03	0.0
2	0.0	-0.9160	-0.5041	0.0	1121.0

Variable Name	Logit Estimate	Standard Errors	t-Statistic
SHINCVC	0.5471	0.2061	2.655
SHINCVNC	0.1701	0.1467	1.160
RMINCV	0.7280	0.9987	0.7290
RINCOME1	0.09354	0.02271	4.118
CDD2	-0.001808	0.0005587	-3.236
A2	4.009	0.3503	11.44

Auxiliary Statistics	At Convergence	At Zero
Log Likelihood	-449.4	-631.5
Percent Correctly Predicted	78.27	50.00

While real income and cooling degree-days are significant and have the expected sign, the coefficients of the inclusive value terms are insignificant in two of three cases. The coefficient estimates on the inclusive value terms are consistent with the hypothesis of random utility maximization.

For comparison we present in Table 3.27 a simple binary logit model of central air conditioning choice that excludes the inclusive values. For the present we argue that either model may be used as a good predictor of the choice of central air conditioning and should perform adequately in the construction of instrumental variables used in the estimation of utilization equations.

Table 3.27

Binary logit model of central air conditioning
choice without inclusive value terms (NIECS)

Variable Name	Logit Estimate	Standard Error	t-Statistic
RINCOME1	.07869	.01273	6.181
CDD2	-.001632	.0001329	-12.28
A2	3.477	.2550	13.64

Auxiliary Statistics	At Convergence	At Zero
Log Likelihood	-460.1	-631.5
Percent Correctly Predicted	77.39	50.00

3.7. The effectiveness of proposed energy policies to influence the selection of household appliance stocks

This section calculates the mean predicted probabilities of HVAC system choice under six alternative levels of building thermal characteristics. The first alternative is the observed dwelling. The second alternative increases existing wall and ceiling insulation to minimum standards proposed by ASHRAE. The third alternative modifies heating and cooling capacities by changing recommended design temperatures.[10] The fourth policy alternative focuses on infiltration losses and recommends that all windows be stormed or double glazed and that simple maintenance reduce the number of air changes by sealing obvious cracks near windows and doors. A fifth alternative to be examined would increase indoor summer temperatures by 5°F and decrease indoor winter temperatures by 5°F. Finally, a sixth alternative combines alternatives two through four to achieve a maximal conservation response. Tables 3.28 and 3.29 summarize the alternative policies for the NIECS and PNW geographic regions.

Our simulation procedure selects for each data set a sample of dwellings of recent vintage. For these dwellings we reestimate predicted capacities and usages under each of the six alternative levels of thermal integrity. The capacities and heating and cooling loads are then used to calculate HVAC capital and operating costs. The procedure by which the output of thermal program under alternative policy scenarios is mapped into capital and operating costs is identical to that used to create

[10]Chapter 2 discusses details of the ASHRAE thermal standards and the connection between design temperature and HVAC capacity.

Table 3.28
Thermal policies for simulation studies (NIECS)

Policy 1 (Insulation)

Minimum Insulation Standards for Walls and Ceilings

	Northeast	North Central	South	West
R-Value Ceiling Insulation	17.14	17.14	19.5	19.5
R-Value Wall Insulation	15.44	15.44	9.45	9.45

Policy 2 (Design Temperatures)

Reduction in Heating and Cooling Design Temperatures

	Northeast	North Central	South	West
Heating Design Temp	12	14	12	14
Cooling Design Temp	7	6	6	5

Policy 3 (Infiltration and Window Glazing)

1) All windows are stormed or double glazed
2) Number of air changes reduced 7 percent

Policy 4 (Thermostat Temperature)

1) Increase indoor summer temperature from 75° F to 80° F
2) Decrease indoor winter temperature from 70° F to 65° F

explanatory variables in the HVAC choice models using the observed sample data.

We have evaluated operating costs using 1978 energy prices and the forecasted values of energy prices for the years 1985, 1990, and 2000. Tables 3.30 and 3.31 summarize the real growth factors in alternative fuels assumed in the simulations.[11] The forecasts indicate that the price of natural gas will nearly double by the year 2000. Electricity and fuel oil are assumed to grow somewhat less rapidly: They experience real growth of 22 percent and 40 percent respectively. The price projections for the Pacific Northwest indicate that electricity will grow most rapidly, at approximately 3.5 percent per year. Natural gas and fuel oil are assumed to grow more slowly, at approximately 2.0 and 1.5 percent per year respectively. Given six alternative levels of thermal integrity and four forecast years, we have defined 24 distinct projections. We employ the HVAC choice model illustrated in Figure 3.1 to forecast the sample

[11]We assume that the real price of capital goods and household demographics remains constant at 1978 levels.

Table 3.29
Thermal policies for simulation studies (PNW)

Policy 1 (Insulation)

Minimum insulation standards for walls and ceilings:
1) R-value ceiling insulation: 19.5
2) R-value wall insulation: 9.45

Policy 2 (Design Temperatures)

Reduction in heating and cooling design temperatures:
1) Heating design temp: 14° F
2) Cooling design temp: 5° F

Policy 3 (Infiltration and Window Glazing)

1) All windows are stormed or double glazed
2) Number of air changes reduced 7 percent

Policy 4

1) Increase indoor summer temperature from 75° F to 80° F
2) Decrease indoor winter temperature from 70° F to 65° F

mean predicted probabilities for six alternative space heating systems without central air conditioning and four systems with central air conditioning. Specifically we assume specification 13 and 14 in Table 3.21 for the NIECS data and specification 5 of Table 3.22 for the PNW data. The availability of natural gas is assumed to remain constant in each scenario.

We graph the sample mean forecast probabilities by HVAC alternative, policy scenario, and forecast year in Figure 3.2. For each HVAC alternative, a corresponding graph in Figure 3.2 provides scale information and an identification of each plotted curve.[12]

In the NIECS data we see an increase in electric forced-air and electric baseboard systems. Conservation policies, however, could decrease the overall share of electric forced-air systems while increasing the share of electric baseboard systems relative to current standards.

Gas systems reveal decreases in overall saturations. Conservation policies appear to reinforce this effect by further reducing predicted market shares. Oil alternatives show a moderate decrease in market share by 1985 while the overall trend is to increase market share slightly from 3

[12]Figure 3.2 presents the 10 NIECS alternatives followed by the six PNW alternatives. The labels for each alternative are easily found in Figure 3.1 or Table 3.17.

Table 3.30
Energy price projections (national averages, NIECS)[a]

	real growth factors			
	1978	1985	1990	2000
Electricity	1.0	1.08	1.13	1.22
Natural Gas	1.0	1.28	1.50	1.94
Oil	1.0	1.15	1.23	1.40

[a]Fuel price projections are from the Department of Energy and the Brookhaven National Laboratory.

Table 3.31
Energy price projections (PNW)[a]

	real growth factors			
	1978	1985	1990	2000
Electricity	1.0	1.31	1.45	1.78
Natural Gas	1.0	1.09	1.18	1.43
Oil	1.0	1.06	1.10	1.33

[a]Fuel price projections are from Cambridge Systematics/West.

percent to 6 percent. Conservation policies will decrease the prevalence of this alternative by the year 2000. This pattern continues to hold in systems with air conditioning. Interestingly, electric heat pumps, which are forecast to enjoy an increasing market share in new construction and conservation policies, will reinforce this pattern.

In the Pacific Northwest electric systems are forecast to gain relative to other fuel types. Electric baseboard systems, however, initially increase and then decrease in market share as electricity prices reach relatively high levels in the year 2000. Gas forced-air systems are forecast to decrease in prevalence, and conservation policies will strengthen this trend. Oil heating systems, which have relatively little market share in the Pacific Northwest, are not revealed to demonstrate large changes in their penetration.

The proposed ASHRAE standards would appear generally to increase the shares of energy-efficient heating and cooling systems. Forecasts for actual energy usage that result under alternative conservation scenarios are reported in Chapter 6.

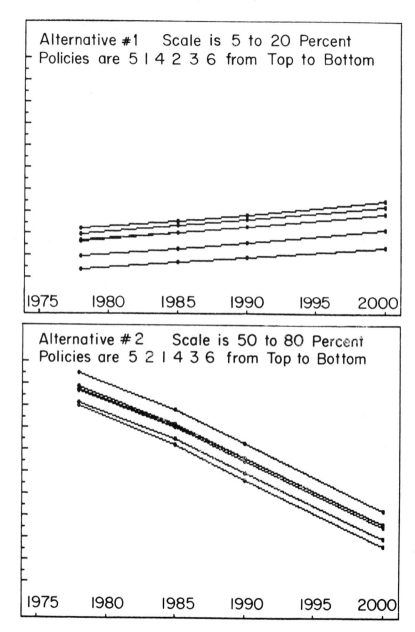

Figure 3.2.1

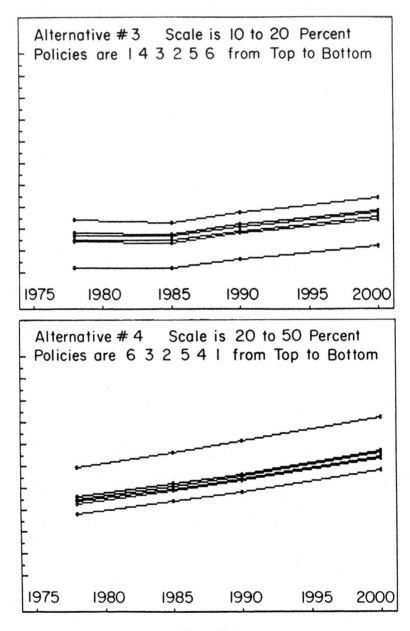

Figure 3.2.2

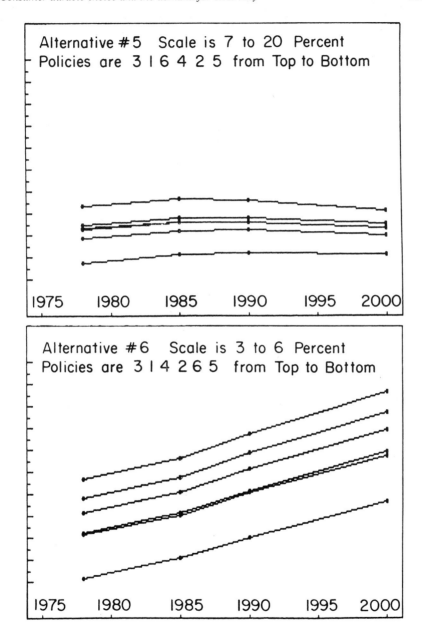

Figure 3.2.3

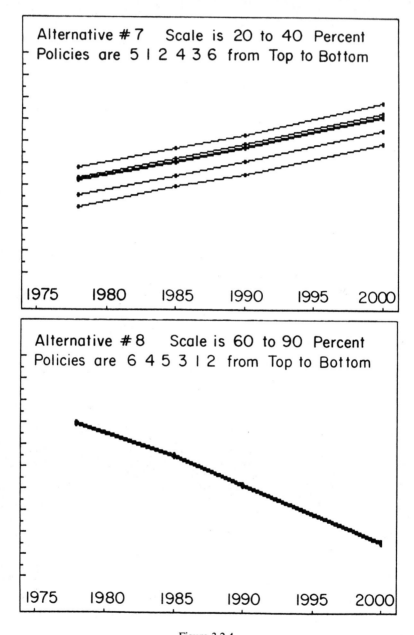

Figure 3.2.4

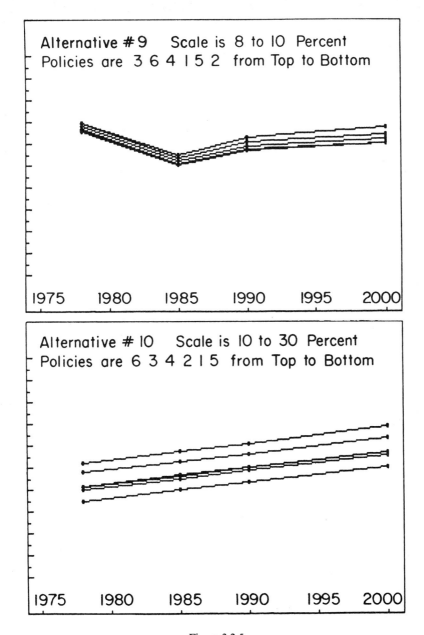

Figure 3.2.5

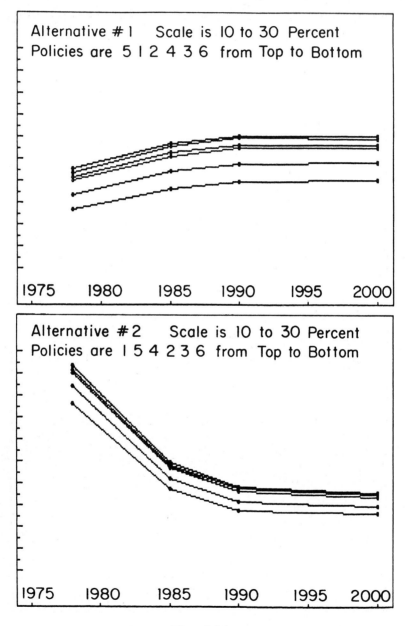

Figure 3.2.6

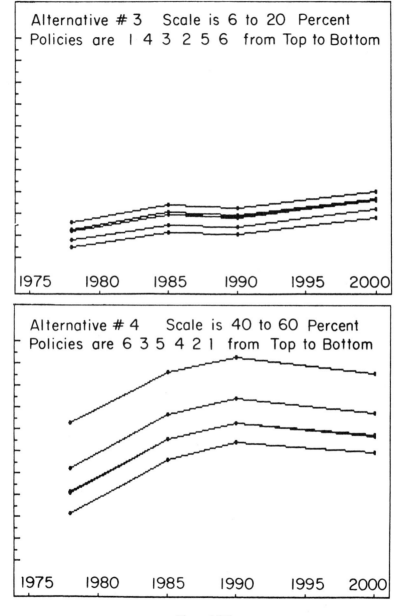

Figure 3.2.7

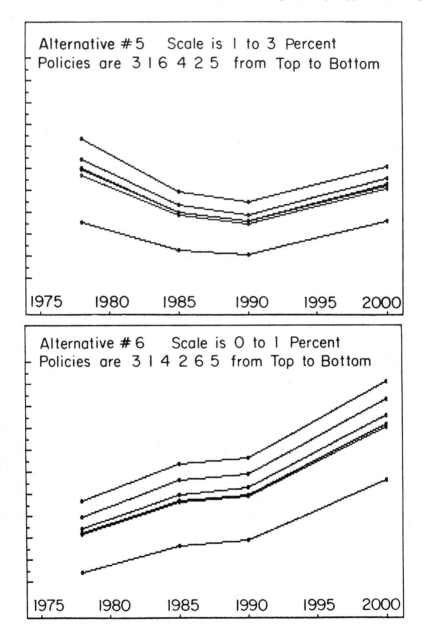

Figure 3.2.8

CHAPTER 4

Rate structure and price specification in the demand for electricity[1]

4.1. Introduction

In this chapter we focus on issues of price specification and construction in conditional demand models. As is well known, electricity has traditionally been priced according to a rate or tariff structure in which the price paid per unit of consumption depends on the level of consumption. Block rate pricing presents interesting theoretical and empirical problems to the classical theory of demand, which assumes parametrically given prices.

The first systematic discussion of price specification in conditional demand models was given by Houthakker (1951a). This was followed by Taylor (1975), who, in a review of the electricity-demand literature, indicates a simple procedure that converts the complex optimization problem of the consumer to the standard case of a linear budget constraint set in marginal prices. The Taylor procedure suggests that regressors in empirical specifications include both marginal and average price. Halvorsen (1982) and Murray (1978) follow this suggestion, while Nordin (1976) and Berndt (1978) indicate corrections and modifications.

More recent studies in the demand for goods facing block pricing indicate that the question of price specification has not been resolved. Billings and Agthe (1980), using the Nordin graphical analysis, demonstrate that the correct price measure is marginal price and that income should be simultaneously adjusted by an appropriate rate structure premium. In this context, the rate structure premium measures the difference between actual expenditure and the cost of consumption priced at marginal cost.

Griffin and Martin (1981) comment that Billings and Agthe fail to account for the endogeneity of price in their empirical work on the demand for water. Foster and Beattie (1981) additionally comment that the distinction between average and marginal costs is inconsequential

---

[1]The author wishes to thank Ernst Berndt, Daniel McFadden, Franklin Fisher, and Jerry Hausman for helpful comments.

empirically. Their empirical evidence is suspect, however, as they fail to implement the test for equality of regressors in two regressions correctly and fail to allow for the endogeneity of prices. Furthermore, a fundamental question of behavior asks whether consumers can detect prevailing marginal rates in the presence of automatic appliances and billing-cycle variations. An alternative hypothesis suggests that consumers respond to a summarizing statistic for the quantity-dependent rate structure such as average price.

Thus several issues remain unresolved in the theory and practice of price specification. In Section 4.2 we review the theory of price specification and consider the comparative static analysis of demand under alternative rate structures. Section 4.3 presents empirical estimates of price responsiveness using cross-sectional individual data from the 1975 Washington Center for Metropolitan Studies (WCMS) survey. We investigate the statistical endogeneity of prices whose construction depends on the observed level of demand. We further investigate the influence of the rate structure premium correction to income and consider the specification error introduced when true marginal price is replaced by the marginal price in the tail end of the rate structure.

Finally, Section 4.4 considers the construction of marginal price when basic observations are limited to total quantity consumed and total expenditure. We then discuss the implications for marginal price construction in the NIECS billing data.

4.2. Specification of price: theory

4.2.1. Quantity-dependent rate structure

We begin by reviewing the general quantity-dependent outlay or expenditure function. Let $B(Q)$ be the total expenditure on electricity when an amount Q is consumed. The rate structure premium, $B(Q) - B'(Q)Q$, is on adjustment to income such that consumers choose quantity level Q at constant marginal price $B'(Q)$.

If $V(P, Y)$ is the indirect utility at prices P and income level Y, then the consumer's optimal choice of quantity subject to the expenditure function $B(Q)$ solves the problem:

$$\max_{Q} V[B'(Q), Y - [B(Q) - B'(Q)Q]]$$

The first-order condition implies that optimal Q is given as the solution to Roy's identity:

$$Q = -\frac{V_P[B'(Q),\ Y - (B(Q) - B'(Q)Q)]}{V_Y[B'(Q),\ Y - (B(Q) - B'(Q)Q)]}$$

$$= D[B'(Q),\ Y - (B(Q) - B'(Q)Q)]$$

where $D[P, Y]$ is the Marshallian demand curve. Thus Q is not generally given as a reduced form in terms of prices, income, or other parameters of the uncompensated demand function. Furthermore, the usual monotonicity properties of the demand function are not sufficient to imply unique optimal consumption levels.

The case of the declining block rate structure is somewhat less complicated and permits us to derive a simple relation among quantity, average price, marginal price, and the rate structure premium. A declining block rate schedule implies an expenditure function that increases in linear segments, the slope of each succeeding segment being smaller than the one preceding it. For $1 < r \leqslant n$:

$$B = C \qquad\qquad\qquad \text{for } 0 \leqslant Q \leqslant X_1$$

$$B = C + \pi_1(Q - X_1) \qquad\qquad \text{for } X_1 < Q \leqslant X_2$$

$$B = C + \sum_{j=1}^{r-1}(X_{j+1} - X_j)\pi_j + \pi_r(Q - X_r) \quad \text{for } X_r < Q \leqslant X_{r+1}$$

where X_i denotes the lower block boundaries and where we have set $X_{n+1} = \infty$. The constant C is the connect charge and π_j is the price of electricity in block j. Suppose measured consumption, Q^*, lies in the rth block so that $X_r < Q^* \leqslant X_{r+1}$ and total expenditure is:

$$B^* = C + \sum_{j=1}^{r-1}(X_{j+1} - X_j)\pi_j + \pi_r(Q^* - X_r)$$

We then define the measured average price as B^*/Q^*, the measured marginal price as π_r, and the rate structure premium (RSP) as the difference between total expenditure and the cost of purchasing the quantity Q^* at the marginal rate π_r: RSP $= B^* - \pi_r Q^*$. Dividing by quantity we obtain the simple relation: average price = marginal price + RSP/Q^*. Nordin (1976) shows that the rate structure premium is an adjustment to income such that consumers choose quantity Q^* at price π_r and income level $Y - $ RSP.

4.2.2. *Comparative static analysis of demand subject to a declining block rate structure*

We now consider the comparative static analysis of demand subject to a declining block rate structure. Let $U[q, Z]$ denote the utility derived from the consumption of electricity q and a Hicksian or numeraire commodity Z. We assume a two-tier tariff for electricity with the price of electricity π given by:

$$\pi = \begin{cases} \pi_1 & \text{for } 0 \leqslant q \leqslant X \\ \pi_2 & \text{for } X < q \text{ and } \pi_1 > \pi_2 \end{cases} \tag{1}$$

If we normalize the price of the numeraire commodity at one, then the budget constraint satisfies:

$$\pi_1 q + Z \leqslant y \qquad\qquad\qquad \text{for } q \leqslant X$$

$$\pi_1 X + (q - X)\pi_2 + Z \leqslant y \qquad \text{for } X < q \tag{2}$$

where y denotes income. We illustrate the declining block tariff and the corresponding budget set in Figure 4.1.

Denote by $D[\pi, y; \beta]$ the Marshallian or uncompensated demand for electricity where β is a vector of behavioral parameters, and let π^* denote the price at which demand equals the lower block boundary, i.e., $D[\pi^*, y; \beta] \equiv X$. Let q_1 denote demand along the segment with slope π_1 and let q_2 denote demand along the segment with slope π_2. Demand along the first budget segment satisfies:

$$q_1 = D[\pi_1, y; \beta] \qquad\qquad \text{for } (\pi_2, \pi_1) \in S_1 \tag{3}$$

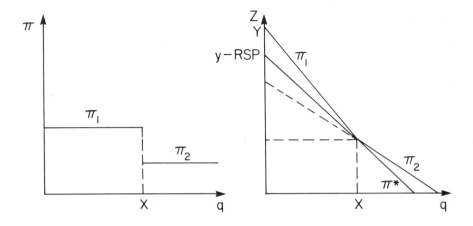

Figure 4.1

while demand in the second segment satisfies:

$$q_2 = D[\pi_2, y - (\pi_1 - \pi_2)X; \beta] \quad \text{for} \quad (\pi_2, \pi_1) \in S_2 \tag{4}$$

We have defined the sets S_1 and S_2 to indicate which price pairs (π_2, π_1) imply optimal first and second segment demand. We show below that S_1 and S_2 constitute a proper partition of all prices that correspond to declining two-part tariffs. Note that the term $(\pi_1 - \pi_2)X$ is the rate structure premium adjustment for demand in the marginal or tail-end block. We now derive certain results concerning local price response.

Lemma 4.2.1. Suppose the uncompensated demand for electricity is decreasing in price and increasing in income. Then:

(a) $\partial q_1/\partial \pi_1 < 0$ for $(\pi_2, \pi_1) \in S_1$

(b) $\partial q_2/\partial \pi_1 < 0$ for $(\pi_2, \pi_1) \in S_2$

(c) $\partial q_2/\partial \pi_2 < 0$ and $(\pi_2, \pi_1) \in S_2$

Proof.

(a) By assumption demand is downward sloping.

(b) $\partial q_2/\partial \pi_1 = (D_y)(-X) < 0$ because we have assumed that electricity is a normal good.

(c) $\partial q_2/\partial \pi_2 = D_\pi + D_y X \leqslant D_\pi + D_y q_2$ because $X \leqslant q_2$.   But $D_\pi + D_y q_2$ equals the partial derivative with respect to price of the Hicksian or compensated demand function (by Slutsky's relation) and is thus negative. Q.E.D.

Remark.   For $\pi_1 \geqslant \pi^*, q_1 \leqslant X$ by   Lemma   4.2.1.   (b).   For $\pi_1 < \pi^*, q_1 > X$ so that optimal demand falls outside the range in which π_1 is the prevailing price. Furthermore Lemma 4.2.1. (c) implies that for $\pi_2 < \pi^*, q_2 > X$. The pattern of prices in which $\pi_2 < \pi^* \leqslant \pi_1$ implies that q_1 and q_2 are each feasible.

Let $V(\pi, y)$ be the indirect utility function corresponding to the problem $\max_{q,Z} U[q, Z]$ subject to $\pi q + Z \leqslant y$. For $\pi_2 < \pi^* \leqslant \pi_1$, the budget segment with price π_1 is optimal when $V(\pi_1, y) > V(\pi_2, y - (\pi_1 - \pi_2)X)$. It is clear that combinations of π_1 and π_2 exist that satisfy $\pi_2 < \pi^* \leqslant \pi_1$ and imply equal indirect utility so that demand for electricity is multivalued. For the set of prices that imply equal indirect utility, a trade-off exists where an increase in π_1 may be compensated by a decrease in π_2. We have the following result:

Lemma 4.2.2. Let:

$$S = \left\{ (\pi_2, \pi_1) \mid V(\pi_1, y) = V(\pi_2, y - (\pi_1 - \pi_2)X) \text{ and } \pi_2 < \pi^* \leqslant \pi_1 \right\}$$

Then $\partial \pi_1/\partial \pi_2 < 0$ for $(\pi_2, \pi_1) \in S$ and for $V_y(\pi_1, y) < V_y[\pi_2, y - (\pi_1 - \pi_2)X]$.

Proof. Let $V_{y_1} = V_y(\pi_1, y)$ and $V_{y_2} = V_y[\pi_2, y - (\pi_1 - \pi_2)X]$.

For $(\pi_2, \pi_1) \in S$:

$$(\partial\pi_1/\partial\pi_2) \cdot V_{\pi_1} = V_{\pi_2} + V_{y_2}[(-X)((\partial\pi_1/\partial\pi_2) - 1)]$$

Then:

$$(\partial\pi_1/\partial\pi_2)(V_{\pi_1} + V_{y_2}X) = V_{\pi_2} + V_{y_2}X$$

which implies

$$(\partial\pi_1/\partial\pi_2) = (V_{\pi_2} + V_{y_2}X)/(V_{\pi_1} + V_{y_2}X)$$

$$= (X - q_2)/(X - q_1(V_{y_1}/V_{y_2})) < 0$$

for $q_1 < X$ and $q_2 > X$ Q.E.D.

To complete the static analysis we need the following result, which indicates the direction of change in indirect utility from changes in price.

Lemma 4.2.3. Let:

$$V_1 = [\pi_1, y] \text{ and } V_2 = V[\pi_2, y - (\pi_1 - \pi_2)X]$$

For $\pi_2 < \pi^* \leqslant \pi_1$:

(a) $\partial V_1/\partial\pi_1 < 0$

(b) $\partial V_2/\partial\pi_1 < 0$

(c) $\partial V_2/\partial\pi_2 < 0$

(d) $\partial(V_2 - V_1)/\partial\pi_1 < 0$ when $V_{y_1} < V_{y_2}$

Proof.

(a) $\partial V_1/\partial \pi_1 = V_\pi(\pi_1, y) < 0$

from the monotonicity property of indirect utility functions.

(b) $\partial V_2/\partial \pi_1 = V_{y_2}(-X) < 0$

(c) $\partial V_2/\partial \pi_2 = V_\pi + V_{y_2}X \leqslant V_\pi + V_{y_2}q_2 < 0$ as $X \leqslant q_2$

(d) $\partial(V_2 - V_1)/\partial \pi_1 = - [V_{\pi_1} + V_{y_2}X]$

$$= - V_{y_2} [X - q_1(V_{y_1}/V_{y_2})] < 0$$

as $V_{y_1}/V_{y_2} < 1$ and $q_1 < X$ Q.E.D.

We collect the results in the following theorem:

Theorem 4.2.1. [Two-Tier Declining Block Rate Comparative Statics]
Let π^* be defined by $D[\pi^*, y; \beta] \equiv X$. Define the functions $\pi_1^*(\pi_2)$ and
$\pi_2^*(\pi_1)$ by:

$$V(\pi_1^*, y) = V(\pi_2, y - (\pi_1^* - \pi_2)X) \qquad \text{and}$$

$$V(\pi_1, y) = V(\pi_2^*, y - (\pi_1 - \pi_2^*)X)$$

respectively. Then equilibrium occurs in the first segment for:

$$S_1 = \left\{ (\pi_2, \pi_1) | \pi^* \leqslant \pi_1 \text{ and } \pi_2^*(\pi_1) \leqslant \pi_2 \leqslant \pi_1 \right\}$$

and equilibrium occurs in the second segment for:

$$S_2 = \left\{ (\pi_2, \pi_1) | 0 \leqslant \pi_2 \leqslant \pi^* \text{ and } \pi_2 \leqslant \pi_1 \leqslant \pi_1^*(\pi_2) \right\}$$

Proof. The shaded region above the diagonal line in Figure 4.2 represents the set of feasible declining block rate structures.

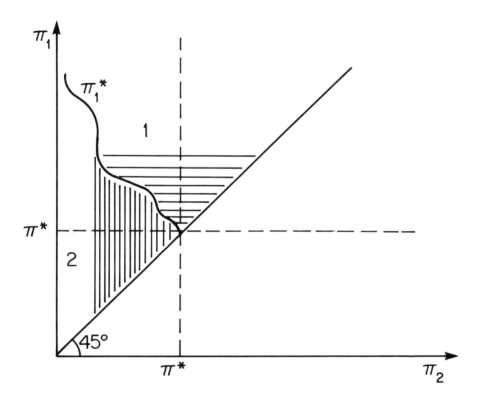

Figure 4.2

The curve with declining slope that intersects the (π^*, π^*) point is the set S of Lemma 4.2.2. Suppose we begin at a point on the curve S and increase π_2 while leaving π_1 unchanged. Because we are in a region in which both budget segments are feasible, Lemma 4.2.3. (c) implies that the increase in π_2 decreases utility V_2. As we began at a point of equal utility and V_2 has decreased while V_1 remains constant, it must be the case that budget segment one is preferred to budget segment two.

Similarly consider a decrease in π_1 leaving π_2 constant. In this case, Lemma 4.2.3. (d) applies so that $V_2 - V_1 > 0$ and budget segment two becomes optimal. In the southwest quadrant above the 45° line, demand occurs in the second budget segment because optimal demand at prices $\pi_1 < \pi^*$ exceeds the block boundary X. The other quadrants are similarly derived using the results of Lemmas 4.2.1 and 4.2.3.   Q.E.D.

Remark. Note that price pairs below the diagonal imply non-decreasing block rate schedules that correspond to convex budget sets. The triangular area in the southwest quadrant below the diagonal implies optimal demand in the second budget segment, while the area below the diagonal in the northeast quadrant implies demand in the first budget segment. The southeast quadrant that includes the boundary $\pi_1 = \pi^*$ but excludes the boundary $\pi_2 = \pi^*$ implies optimal demand at the block boundary X. We further note that the set S of equal utility points has measure zero in the price space of Figure 4.2.

We now use Figure 4.2 to answer simple comparative static problems. Suppose for example that we increase the lower block boundary from X to X'. Figure 4.3 illustrates that the partition moves to an intersection with the 45° line at the point $(\pi^{*\prime}, \pi^{*\prime})$ with $\pi^{*\prime} < \pi^*$. If equilibrium had initially occurred at point A, the discontinuous change in lower block boundary from X to X' would now imply that point A corresponds to optimal demand in budget segment one versus the initial equilibrium in budget segment two.

Finally we note that our comparative static analysis applies to the more general case of multiple-tier declining block rate schedules where we interpret π_2 as the marginal rate and π_1 as the intramarginal average price, i.e., the average price up to but not including the marginal block.

4.3. Specification of price: empirical results

We now address the issue of price specification with an econometric analysis of the 1975 survey of 1502 households carried out by the Washington Center for Metropolitan Studies (WCMS) for the Federal Energy Administration. Individual household locations (identified at the level of primary sampling units) permitted matching of actual rate schedules used in 1975 to each household. The use of disaggregated data is

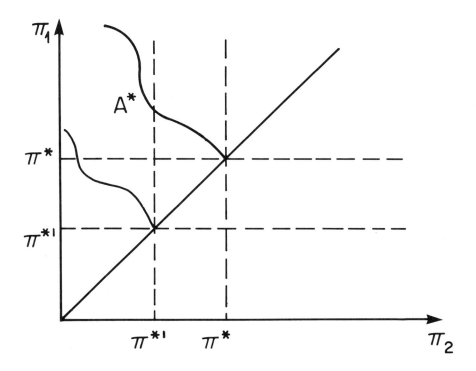

Figure 4.3

necessary to avoid the confounding effects of misspecification aggregation bias or approximation of the rate data.

We resolve four empirical issues related to the estimation of the demand for electricity: (1) measured average price and measured marginal price are statistically endogenous so that least squares techniques are not appropriate for the determination of price elasticities, (2) while the rate structure premium adjustment has established theoretical merit, its statistical contribution is negligible, (3) consumer behavior in the demand for electricity follows the marginal price specification rather than the average price specification, and (4) estimates of price responsiveness are not statistically different using the tail-end price rather than the true marginal rate.

4.3.1. *Endogeneity of measured prices*[2]

The general proposition is that explanatory variables that utilize the observed consumption level introduce correlation between those variables and the error term. To illustrate the direction of least squares estimation bias, write the demand for electricity equation as $Q = \beta p + Z\delta + \varepsilon$ where p is the measured marginal price with coefficient β, Z is a vector of socioeconomic variables with coefficient vector δ, and ε is the equation error. For simplicity assume that p is not correlated with Z so that $\hat{\beta}_{LS} = \beta + p'\varepsilon / p'p$. An unobserved increase in electricity consumption induces a decrease in price so that we expect an *a priori* negative correlation between p and ε. The formula for $\hat{\beta}_{LS}$ shows that least squares overestimates in absolute magnitude the price-response coefficient β.[3]

McFadden et al. (1978) and Hausman et al. (1979) have demonstrated that an instrumental variable estimation technique provides consistent estimates of the electricity demand equation where instruments are constructed utilizing predicted rather than actual consumption to determine measured prices. In forming predicted consumption levels, all endogenous variables are purged from the set of explanatory variables. One must ensure that the instruments so constructed are not exact linear combinations of the exogenous variables included in the demand for electricity equation. This is usually not a problem given the non-linearity of the rate schedule and the existence of other prices that are exogenous. The tail-end block price, for example, will be used in exactly this role.[4]

To establish empirical verification of the hypothesis of endogeneity of measured price, we apply the specification test proposed by Wu (1973) and recently discussed in Hausman (1978). The methodology consists of isolating a group of explanatory variables whose endogeneity is under

---

[2]A source of bias not discussed in this chapter arises from the endogeneity of appliance ownership dummies. Generally, unobserved factors that influence the choice of a durable will also influence its use. For a complete discussion of this problem, see Dubin and McFadden (1984), who find evidence that this leads to underestimates (in magnitude) of the true price effects.

[3]This result is further true when p is correlated with Z. It is not in general possible, however, to determine the magnitude of the bias when several explanatory variables are correlated with the error term.

[4]Recently Terza and Welch (1982) have considered a correction like that of Heckman (1979) for price endogeneity. Their method had been proposed earlier in a footnote to Burtless and Hausman (1978) and requires a maximum likelihood estimation to predict the quantity price pair that yields maximum utility. Here we favor the simpler instrumental variable technique to correct for endogeneity.

test. Using the result that the least squares estimator has zero asymptotic covariance with its difference from the instrumental variable estimator, we are able to form a simple statistic that is asymptotically chi-squared under the null hypothesis of statistical exogeneity for the test group.

To illustrate the test, write the demand for electricity in schematic form as $Q = X\beta + Z\gamma + \varepsilon$ where X is a k-vector of price and income terms under various specifications and Z is a group of assumed exogenous variables. The variables in X are presumed to be suspect of endogeneity. The test statistic is then:

$$T = (\hat{\beta}_{IV} - \hat{\beta}_{LS})'[V[\hat{\beta}_{IV}] - V[\hat{\beta}_{LS}]]^{-1}(\hat{\beta}_{IV} - \hat{\beta}_{LS}) \overset{A}{\sim} \chi^2(k)$$

where $\hat{V}$ is the estimated variance covariance matrix and k is the number of coefficients in β.

The dependent variable in each estimated equation is monthly consumption of kilowatt-hours of electricity used by the family in 1975. The socioeconomic variables include appliance ownership dummies for the electric dishwasher, electric washing machine, food freezer, electric range, color television, black-and-white television, electric clothes dryer, and central air conditioner. To capture the effects of climate, the annual number of cooling degree-days (the number of days in which the daily average temperature was greater than 65°) and this number multiplied by the central air conditioner dummy and by the number of room air conditioners, respectively, were included, as well as scale variables for the number of rooms, the number of persons, and the number of room air conditioners.[5]

Price terms included the average price, measured marginal price, and the tail-end block rate. These rates are used below in various combinations and are taken from the rate schedules prevailing in the winter of 1975. In Table 4.1 we present the mean values of all variables. To demonstrate the bias induced by least squares under the marginal price specification, we compare the least squares and instrumental variable estimates of the equation: $Q = \alpha(\text{measured marginal price}) + Z\delta + \varepsilon$. For brevity we report in Table 4.2 the coefficient estimates of the variables: measured marginal price, income, electric water heating, and electric space heating. At sample means the price elasticity implied by least

---

[5]The empirical results appear to be robust with respect to choice of exogenous variables and alternative functional specifications.

Table 4.1

Mean values for variables in the demand models

Variable Name[a]	Description	Mean
AKWH75	Monthly consumption of electricity in 1975	916.5
RATE	Measured marginal price in 1975	.02427
AVPRICE	Measured average price in 1975	.03128
WMPE75	Winter tail-end block price for electricity in 1975	.02138
INCOME	Monthly income of household head	1322.
RSP	Measured rate structure premium	5.151
WHE	Electric water heating dummy	0.2728
SHE	Electric space heating dummy	0.1411
ROOMS	Number of rooms in household	6.078
PERSONS	Number of persons in household	3.550
CAC	Central air conditioning dummy	0.2890
CDDCAC	(Annual cooling degree-days) * (CAC)	463.7
RACNUM	Number of room air conditioners	.4382
CDDRACNUM	(Annual cooling degree-days) * (RACNUM)	642.3
AUTOWSH	Automatic washing machine dummy	0.8898
AUTODSH	Automatic dishwasher dummy	0.4921
FOODFRZ	Food freezer dummy	0.5323
ELECRNGE	Electric range dummy	0.6411
ECLTHDR	Electric clothes dryer dummy	0.4990
BWTV	Black-and-white television dummy	0.5806
CLRTV	Color television dummy	0.7446

[a]A subsample of 744 of the original 1502 observations was selected so that all price and income data were positive and so that complete information was available for each individual.

squares is -0.266 while the instrumental variable estimates imply a price elasticity of -0.159. The direction of the bias agrees with our *a priori* expectation that least squares will overestimate in magnitude the price sensitivity coefficient.

Taylor (1975) reports both short-run and long-run price and income elasticities. Of nine estimates of residential elasticities, two used marginal price. Each of the studies by Houthakker (1951a, 1951b) reports short-run elasticities of approximately -0.90.[6] Both our least squares and instrumental estimates are well below this estimate in magnitude, but are entirely consistent with other estimates of electricity demand price elasticity using an average price specification.[7]

[6]The rate schedule in Houthakker's study consisted of a connect charge and a fixed marginal price. The marginal price elasticity estimated by Houthakker is not tainted by simultaneity bias.

[7]Studies by Acton, Mitchell, and Mowill (1976) and Taylor, Blattenberger, and Verleger (1977) find short-run price elasticities from -.08 to -.35 with endogenous marginal price specifications.

Table 4.2
Measured marginal price model

Variable[a]	LS Estimates	IV Estimates
Measured marginal price	-10050.	-6006.
	(-5.909)	(-3.269)
Income	.08169	.07570
	(3.330)	(3.071)
Electric water heating	405.6	404.5
	(10.22)	(10.15)
Electric space heating	694.8	714.9
	(14.8)	(14.40)
R^2	.7074	.7051
Number of observations	744	744
Sum of squared residuals	.9094E+8	.9166E+8
Standard error of regression	354.2	355.6

[a]In Tables 4.2-4.6 coefficient estimates are *not* reported for these variables: PERSONS, BWTV, ROOMS, RMCLCAC, CDDCAC, CAC, RACNUM, CDDRACNUM, FOODFRZ, ELECRNGE, CLRTV, ECLTHDR, AUTODSH, AUTOWASH, and the intercept. The dependent variable is AKWH75. The t-statistics are presented in parentheses.

The Hausman statistic for the endogeneity test of measured marginal price is computed to be 34.18. This well exceeds the critical value for a chi-squared test of any size given the single degree of freedom. We note that the respective income elasticities for least squares and instrumental variables are 0.118 and 0.109. Both estimates are consistent with those obtained in previous studies.

If the same test is repeated using measured average price in place of measured marginal price, we find price elasticities for least squares and instrumental variables of -0.437 and -0.416 respectively. Note that the direction of bias is the same as that obtained with measured marginal price —a general increase in price sensitivity magnitude. Income elasticities were robustly estimated at 0.120 and 0.104 for the two procedures. The chi-squared statistic was computed in this case to be 118.2, which well exceeds the critical value of 3.84 for a 5-percent test. Parameter estimates for the average price specification are reported in Table 4.3.

In summary, we remark that previous studies in the demand for electricity have undoubtedly been subject to the bias illustrated above. *The bias has been demonstrated to be statistically significant for the two most common specifications of price and is qualitatively impressive on the order of 67 percent.*[8]

[8]The bias for the average price specification is not as large at approximately 5 percent.

Table 4.3
Measured average price model

Variable	LS Estimates	IV Estimates
Average price	-12810.	-4266.
	(-8.731)	(-2.563)
Income	.08304	.07221
	(3.484)	(2.959)
Electric water heating	388.8	398.1
	(10.05)	(10.06)
Electric space heating	669.2	719.2
	(13.90)	(14.45)
R^2	.7225	.7095
Number of observations	744	744
Sum of squared residuals	.8626E+8	.9029E+8
Standard error of regression	344.9	352.9

4.3.2. Rate structure premium adjustment

From Table 4.1 we see that the mean value of rate structure premium is $5.15 compared with the mean value of income of $1322/month. The negligible value of RSP as compared with INCOME implies that the difference (INCOME - RSP) could not be distinguished from general measurement error in the definition of monthly income. In Table 4.4 we present instrumental variable estimates of the electricity demand equation using the marginal price specification and income adjusted by the rate structure premium.

Comparison of the estimates in Table 4.4 with estimates given in Table 4.2 for instrumental variables demonstrates the qualitative similarity. Based on these results *we do not advocate the rate structure premium correction to income in the WCMS data for 1978.* This confirms the findings of Hausman et al. (1979) for insignificance of the RSP adjustment.

4.3.3. Average versus marginal price

Estimation in demand for electricity studies has followed the predominant usage of either marginal or average price. A simple observation will allow us to nest both the marginal and average price specification in a more general model. We have demonstrated above that the difference between measured average price and measured marginal price is the rate

Table 4.4
Measured marginal price and adjusted income

Variable	IV Estimates
Measured marginal price	-6006.
	(-3.269)
Adjusted income	0.07560
	(3.067)
Electric water heating	404.5
	(10.15)
Electric space heating	715.0
	(14.40)
R^2	.7050
Number of observations	744
Sum of squared residuals	.9167E+8
Standard error of regression	355.6

structure premium divided by measured consumption. Hence an unrestricted specification of marginal and average prices has the form: $Q =$ (measured marginal price)α_0 + (rate structure premium / quantity)α_1 + $Z\delta$ + ε. Clearly when α_0 equals α_1 we have the average price specification. When $\alpha_1 = 0$ we have the marginal price specification. Ordinary least squares and instrumental variable estimates for the unrestricted model are presented in Table 4.5.

Table 4.5
Average versus marginal price

Variable	LS Estimates	IV Estimates
Measured marginal rate	-10130.	-6430.
	(-6.158)	(-3.352)
Rate structure premium/quantity	-22410	10040.
	(-7.236)	(1.777)
Income	.07702	.07846
	(3.248)	(3.068)
Electric water heating	374.9	418.4
	(9.717)	(9.961)
Electric space heating	673.6	722.1
	(14.10)	(14.00)
R^2	.7272	.6840
Number of observations	744	744
Sum of squared residuals	.8481E+8	.9823E+8
Standard error of regression	342.3	368.3

For brevity we report only the coefficient estimates of measured marginal price, rate structure premium/quantity, income, WHE, and SHE. The Hausman statistic of 83.8 with the two degrees of freedom confirms the endogeneity of the explanatory variables measured marginal price and rate structure premium/quantity.

Using the instrumental variables estimates in Table 4.5, we compute a Wald test of the hypothesis that the coefficients of measured marginal rate and rate structure premium/quantity are equal. The test statistic that compares the difference in the estimated coefficients has a value of 7.09 and is distributed chi-squared with one degree of freedom (the number of imposed restrictions). We thus reject the average price specification at the 1-percent critical level. Furthermore, the individual t-statistics for the coefficients of measured marginal price and RSP/Q confirm the marginal price specification as the former coefficient is significant while the latter is insignificant at the 5-percent level.[9] It is interesting to note that inspection of the least squares estimates would lead one to choose the average price specification over the marginal price specification. Given the differential in sum of squared residuals for the measured marginal price and average price specifications (using the consistent estimates in Tables 4.2 and 4.3 respectively) it is likely that a non-nested test (see Pesaran (1975), for example) would also discriminate between the two models. We are thus led to conclude that *consumer behavior in the demand for electricity follows the marginal price specification rather than the average price specification.*

4.3.4. *Measurement error in marginal price*

We now consider the impact of the measurement-error misspecification that results from the use of the tail-end rate in place of the measured marginal rate. In Table 4.6 we reproduce the least squares regression results for this specification. Note that least squares estimation provides consistent parameter estimates because the tail-end price is by definition exogenous. The use of the tail-end rate in place of the measured marginal rate introduces measurement error in the price variable. It is not

[9]We have rejected the null hypothesis that demand for electricity follows the average price specification. This of course is not equivalent to accepting the marginal price specification. Given the sign change on the coefficient of RSP/Q and its standard error, we cannot reject the marginal price specification.

Table 4.6
Tail-end price

Variable	LS Estimates
Tail-end price	-6828.
	(-3.644)
Income	.08299
	(3.277)
Electric water heating	414.1
	(10.26)
Electric space heating	721.7
	(14.51)
R^2	.6988
Number of observations	744
Sum of squared residuals	.9361E+8
Standard error of regression	359.3

appropriate to apply the usual measurement error bias formulae, however, because price is expected to reveal significant correlation with the other explanatory variables and because the difference between the two measures of price is not a mean zero random disturbance.

Comparing the estimate of the tail-end price coefficient in Table 4.6 with the consistent estimate of the measured marginal price coefficient in Table 4.2, we see that relative to the standard error the difference is not significant: (t = ((-6006.) - (-6828.))/1837. = 0.45). This is confirmed through the inspection of the variables WMPE75 and RATE; the correlation coefficient between the two variables in 0.87 and the mean difference is approximately one-third of a standard deviation. While there is no specific suggestion that the rate schedules in the WCMS data are flat, these estimates suggest that *many individuals are close to the tail end of the rate schedule so that measured marginal rates are well approximated by the tail-end price.*[10]

---

[10]Using the results of Section 4.2, we see that RSP = $X(\pi_1 - \pi_2)$. As average price less marginal price equals RSP/Q, we have average price = $\pi_2 + (X/Q)(\pi_1 - \pi_2)$. For $Q \gg X$, average price and marginal price will be approximately equal. Thus individuals with demand much larger than the tail-end block will have approximately equal marginal, average, and tail-end prices.

4.4. Measurement of price: theory and estimation

This section investigates the construction of marginal price when basic observations are limited to total quantity consumed and total expenditure. We begin with an analysis of eight locations from the WCMS (1975) data set for which precise matching of rate schedules to households was possible. We compare the two-part tariff approximation to the actual rate schedule and attempt to illustrate the qualitative and quantitative bias in each physical location. We then examine seven locations from the National Interim Energy Consumption Survey (NIECS, 1978) for which only total expenditure and quantities are observed by billing periods. Under the assumption that households within a primary sampling unit are served by a common utility, we attempt to distinguish between all electric and seasonal rates.

In Figures 4.4-4.11, we plot expenditure versus quantities for eight WCMS households. The figures are organized in pairs: (1) the plot of expenditures versus quantities, and (2) the plot of the prevailing rate schedule. A solid circle denotes points chosen from the rate schedule while a solid triangle and an open circle indicate one and two observations respectively. For each location we give the estimates of the two-part tariff approximations, the actual tail-end price, and the appropriate connect charge.

Figure 4.4.1 illustrates that nine of the ten observations from Boston, Massachusetts, correspond to the tail-end price. The estimate of marginal price from the two-part tariff approximation is 0.0373, while the actual tail-end rate is 0.03693. The standard error of the slope estimate is 0.000215 so that a t-test for significance of the difference is rejected at the 5 percent level: (one-sided test, degrees of freedom = 8, size = 5 percent, t = 1.172). The situation in Figures 4.5.1 and 4.5.2 is qualitatively similar. In this case nine of the 11 observations lie in the tail end of the rate structure. The estimate of the slope is 0.0317, which is again not statistically different from the tail-end price 0.03169 (one-sided test, degrees of freedom = 9, size = 5 percent).

In Figure 4.6 fewer observations are in the tail. The estimate of the slope coefficient is 0.0217 while the tail-end price is 0.02043. The t-statistic for the difference is 3.86, which is significant for a one-sided 5-percent test given the 12 degrees of freedom. Figure 4.7 illustrates a near-perfect fit as the underlying rate schedule is flat. By contrast the distribution of points in Figure 4.8.1 suggests that the two-part tariff

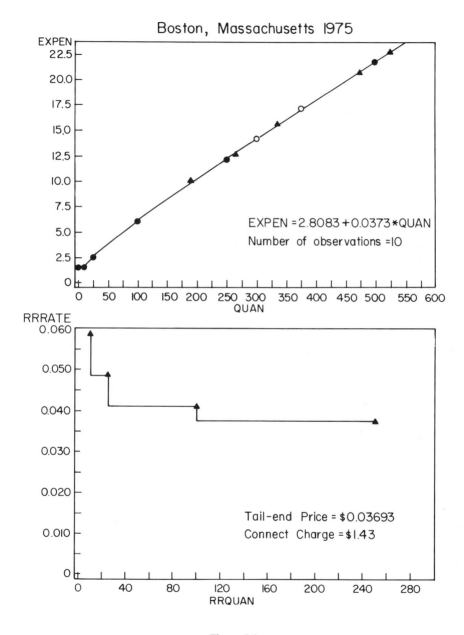

Figure 4.4

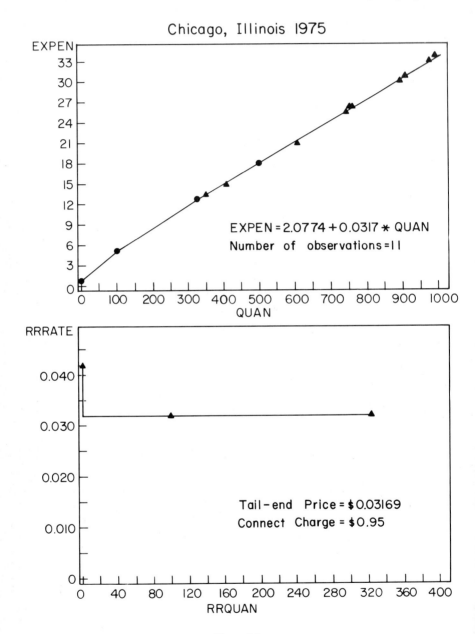

Figure 4.5

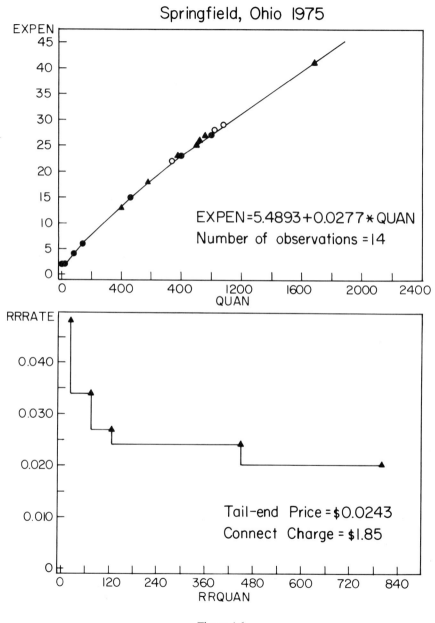

Figure 4.6

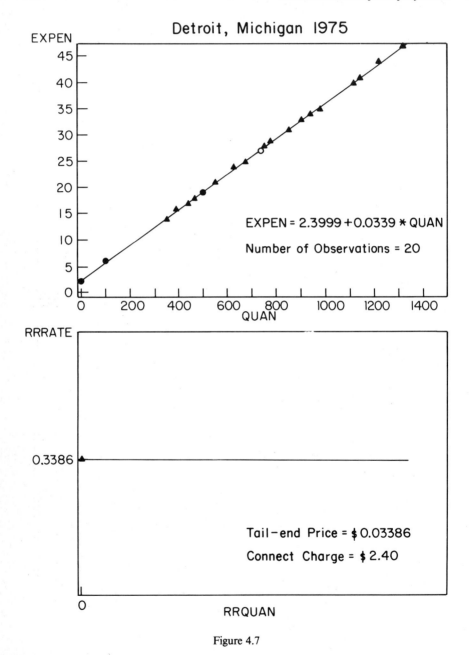

Figure 4.7

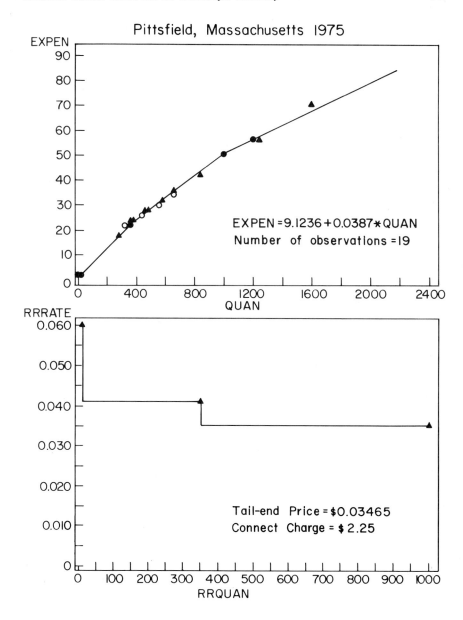

Figure 4.8

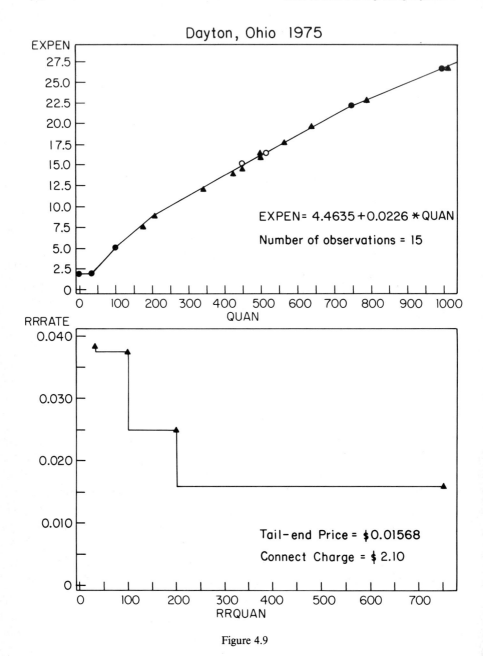

Figure 4.9

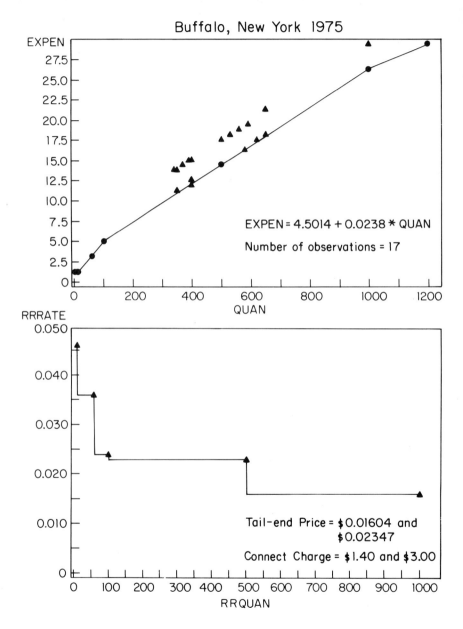

Figure 4.10

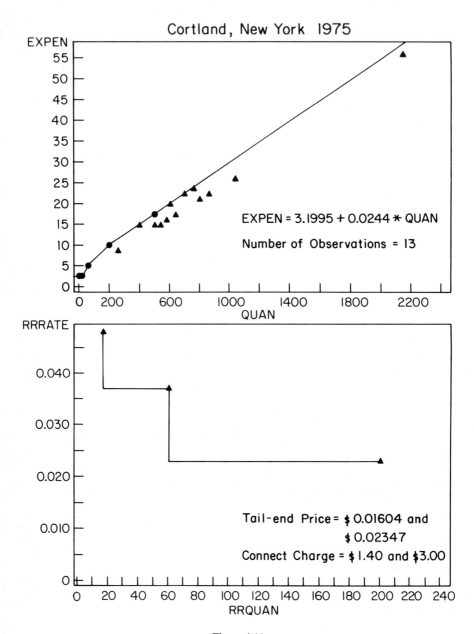

Figure 4.11

should not approximate the declining block rate schedule very accurately. In this case the estimated slope coefficient is 0.0387 while the true tail-end rate is 0.03465. With 17 degrees of freedom, we reject the hypothesis that the estimate of the tail-end rate and the actual rate are equal (t = 9.01, size = 5 percent, degrees of freedom = 17). Figure 4.9 is qualitatively similar to Figure 4.8: (t = 11.58 with 13 degrees of freedom).

Figures 4.10 and 4.11 illustrate a quite different phenomenon. Clearly two separate rate schedules were operative for Buffalo and Cortland, New York, in 1975. These rates are given in Figures 4.10.2 and 4.11.2 respectively. In this case it is seen that the two-part approximation to multiple rate schedules does not properly discriminate the estimated from actual rates. The estimated rate in Figure 4.10 is significantly different from one marginal rate with t = 3.52, yet is not significantly different from the other marginal rate with t = 0.17 (corresponding values are t = 8.64 and t = 0.92 for Figure 4.11).

In summary we see that *the two-part tariff approximation to the declining rate schedule works quite well when many observations lie in the tail-end block.* When more than the one rate schedule prevails within a primary sampling unit, however, it is possible to incorrectly estimate the tail price. As the eight WCMS locations are not necessarily representative of the complete sample, it is not possible to make a statement about general misspecification from only their analysis.

The following calculation attempts to bound the estimation error inherent in the use of a two-part tariff approximation. Essentially, misspecification arises because the rate structure premium varies with quantity. If the rate structure premium were constant, then the rate structure would be exactly in two-part tariff form. We thus apply a simple misspecification argument to estimate the bias.

Recall that by definition: expenditure = rate structure premium + marginal price · quantity. For household i ($i=1, 2, ..., T$), we write total expenditure, $EXPEN_i$, as: $RSP_i + \beta q_i + \varepsilon_i$ where: RSP_i = rate structure premium for household i, β = marginal rate, q_i = quantity consumed by household i, and ε_i = the disturbance term. Rewrite the true model as:

$$EXPEN_i = \alpha + \beta q_i + v_i$$

where:

$$v_i = \varepsilon_i + RSP_i - \alpha$$

Least-square estimation implies:

$$\hat{\beta} - \beta = \frac{\sum\limits_{i} (q_i - \bar{q})(v_i - \bar{v})}{\sum\limits_{i} (q_i - \bar{q})^2}$$

Because $v_i - \bar{v} = (\varepsilon_i - \bar{\varepsilon}) + (RSP_i - \overline{RSP})$ we have:

$$\hat{\beta} - \beta = \frac{T^{-1} \sum (q_i - \bar{q})(\varepsilon_i - \bar{\varepsilon})}{T^{-1} \sum (q_i - \bar{q})^2}$$

$$+ \frac{T^{-1} \sum (q_i - \bar{q})(RSP_i - \overline{RSP})}{T^{-1} \sum (q_i - \bar{q})^2}$$

so that $Plim_{T \to \infty} (\hat{\beta} - \beta) = (\sigma_{RSP}/\sigma_q) \cdot \text{correl}(q, RSP)$ where σ_{RSP} and σ_q are the sample standard deviations of RSP and quantity respectively and correl(q, RSP) is the sample correlation coefficient between quantity and rate structure premium.

In the WCMS data the correlation of rate structure premium and quantity is 0.4659, while the standard deviation of rate structure premium and quantity are 2.906 and 646.3 respectively. Hence the two-part tariff approximation bias underestimates the true marginal rate by 0.002095. Using these estimates, the two-part approximation would imply marginal price of 0.02348 relative to the mean value tail-end price of 0.02138. This difference is about 25 percent of one standard deviation in the tail-end price.

In summary, it appears that the two-part tariff approximation adequately represents the declining block rate schedule in the determination of the tail-end block rate for WCMS households in 1975. As the estimated marginal prices are independent of quantity consumed, they may be treated as exogenous in demand equations. The results in Section 4.3.4 lead us to conclude that marginal rates so determined should adequately measure the effects of the endogenous measured marginal rate.[11]

[11] Billings (1982) also employs the two-part tariff approximation in the determination of marginal price. His analysis does not provide an estimate of possible measurement error.

Based on these considerations, we employ the two-part tariff approximation to determine marginal rates for NIECS households. A simple dummy variable technique permits separate estimates of seasonal and all-electric rates. Details of price construction are given in Appendix A.

CHAPTER 5

Two-stage estimation methods for the switching regime model with known regimes

5.1. Introduction

This chapter considers the switching regression model with known regimes. The primary focus is to utilize distributional assumptions and compatible estimation techniques that permit the modeling of many regimes simultaneously. Additionally we compare three estimation techniques for relative asymptotic efficiency and derive the relevant limiting distributions in each case. The framework developed here is relevant in the estimation of wage-earnings functions conditional on occupational choice or, more to the point, in the determination of energy usage conditional on the selection of an appliance stock.

In either of the two examples offered, a large number of qualitative outcomes limit the empirical implementation of the multinomial probit model. The further limits imposed by the joint normality assumption in the discrete and continuous equations have led Dubin and McFadden (1984) to consider in some detail the multinomial logit model. Along similar lines, Olson (1980) considers the uniform distribution for the latent variable equation that yields the linear probability specification. Hay (1980) has used the multinomial logit model to study wage earnings conditional on occupational choice.

A few notes on the place of this work in relation to that of other investigators seem relevant. The switching regime model was introduced in its present form by Quant (1972) and Goldfeld and Quant (1973). The most comprehensive treatment of the switching regression is given by Hartley (1977). His analysis considers maximum likelihood estimation under the assumption of joint normality for the six subcases of switching regression models. These are in turn:

1) the complete information system with uncorrelated disturbances,

2)   the complete information system with correlated disturbances,

3)   the known regime system with uncorrelated disturbances,

4)   the known regime system with correlated disturbances,

5)   the unknown regime system with uncorrelated disturbances,

6)   the unknown regime system with correlated disturbances.

"Complete information" refers to the ability to observe the latent variables in the switching equation. "Correlated disturbances" refers to the statistical dependence of the errors in the main equation with the errors in the latent variable switching equation. Complete information systems can be regarded as a special case of seemingly unrelated regressions. The known regime case can be analyzed by two-step consistent methods or by maximum likelihood techniques. The unknown regime cases seem to be amenable to only the latter technique.

This section considers only case four above. Related work on this subcase is the paper by Lee and Trost (1978) in which maximum likelihood and the Amemiya-Heckman correction methods are employed under an assumption of normality. The papers by Olson and Hay, mentioned above, only use the Amemiya-Heckman methods. More recently, Lee (1982) generalizes bivariate distributions by mixing densities through a translation method discussed in Mardia (1970).

5.2. Estimation of the switching regime model with known regimes

We begin with the linear in parameters form:

$$y_t = \delta_t X_{1t}\beta_1 + (1 - \delta_t)X_{2t}\beta_2 + \eta_t, \quad t = 1, 2, ..., T \tag{1}$$

where β_j is column vector of K_j parameters, X_{jt} is row vector of K_j explanatory variables, y_t is a scalar dependent variable, η_t is a scalar equation error, and δ_t is a scalar dummy variable. We assume that δ_t equals one when the latent random variable y_t^* is less than zero. Equation (1) and the stochastic specification for y_t^* form a dummy endogenous simultaneous equation system.

We now consider several two-step estimation procedures that provide consistent estimates of the parameters β under the assumption that the dummy indicator variable is endogenous. We define:

$$W(\delta) = <\delta_t X_{1t}, (1 - \delta_t)X_{2t}> \tag{2}$$

$$W(p) = <p_t X_{1t}, (1 - p_t)X_{2t}> \tag{3}$$

$$W(\hat{p}) = <\hat{p}_t X_{1t}, (1 - \hat{p}_t)X_{2t}> \tag{4}$$

The order of the matrices $W(\delta)$, $W(p)$, and $W(\hat{p})$ is $T \times (K_1 + K_2)$. The matrix $W(p)$ is constructed by replacing the indicator δ_t in $W(\delta)$ by its expected value denoted p_t. The matrix $W(\hat{p})$ is constructed by replacing the indicator δ_t in $W(\delta)$ by a consistent estimate of the true probability denoted $\hat{p}_t$.

Define two least squares projections:

$$W \equiv W(p)[W(p)'W(p)]^{-1}W(p)'W(\delta)$$

$$\hat{W} \equiv W(\hat{p})[W(\hat{p})'W(\hat{p})]^{-1}W(\hat{p})'W(\delta) \tag{5}$$

We express equation (1) alternately as:

$$y = W(\delta)\beta + v^0 \qquad\qquad v^0 \equiv \eta \tag{6}$$

$$y = W\beta + \xi^1 \qquad\qquad \xi^1 \equiv \eta + [\, W(\delta) - W\,]\beta \tag{7}$$

$$y = \hat{W}\beta + \xi^2 \qquad\qquad \xi^2 \equiv \eta + [\, W(\delta) - \hat{W}\,]\beta \tag{8}$$

$$y = W(p)\beta + v^3 \qquad\qquad v^3 \equiv \eta + [\, W(\delta) - W(p)\,]\beta \tag{9}$$

$$y = W(\hat{p})\beta + v^4$$

$$v^4 \equiv \eta + [W(\delta) - W(p)]\beta - [W(\hat{p}) - W(p)]\beta \tag{10}$$

In the presence of correlation between δ_t and η_t, ordinary least squares applied to (6) yields inconsistent estimates of β. Consider in turn the least squares estimators of equations (7) to (10).

$$\hat{\beta}^1 = (W'W)^{-1}(W'y) \tag{11}$$

$$\hat{\beta}^2 = (\hat{W}'\hat{W})^{-1}\hat{W}'y \tag{12}$$

$$\hat{\beta}^3 = (W(p)'W(p))^{-1}W(p)'y \tag{13}$$

$$\hat{\beta}^4 = (W(\hat{p})'W(\hat{p}))^{-1}W(\hat{p})'y \tag{14}$$

Note that $\hat{\beta}^1$ and $\hat{\beta}^2$ depend on the unobserved parameter p_t and are therefore nonviable estimators of β. We include these estimators to study the effect of efficiency loss due to prior estimation of the parameter p_t. Observe that $\hat{\beta}^1$ and $\hat{\beta}^2$ are instrumental variable estimators. In the first stage of (11), the endogenous variables $W(\delta)$ are projected onto the exogenous set of instruments $W(p)$. The resultant instrument matrix is given by W. In the second stage, we find:

$$(W'W(\delta))^{-1}W'y = (W'W)^{-1}W'y = \hat{\beta}^1.$$

Moreover, note that:

$$\hat{\beta}^1 - \beta = (W'W(\delta))^{-1}W'\xi^1 = (W'W)^{-1}W'\eta \quad \text{as}$$

$$W'\xi^1 = W'[\eta + (W(\delta) - W)\beta] = W'\eta$$

where we use the fact that the residual portion of ξ^1, $(W(\delta) - W)\beta$, is orthogonal to W. These comments apply directly to $\hat{\beta}^2$ and produce the instrumental variable estimator

$$\hat{\beta}^2 - \beta = (\hat{W}'W(\delta))^{-1}\hat{W}'\eta.$$

The estimators $\hat{\beta}^3$ and $\hat{\beta}^4$ satisfy

$$\hat{\beta}^3 - \beta = (W(p)'W(p))^{-1}W(p)'v^3 \quad \text{and}$$

$$\hat{\beta}^4 - \beta = (W(\hat{p})'W(\hat{p}))^{-1}W(\hat{p})'v^4 \quad \text{respectively.}$$

It is important to note that the residual portions of v^3 and v^4 are not orthogonal to $W(p)$ or $W(\hat{p})$ except asymptotically. As $W(p)'v^3 \neq W(p)'\eta$, it is not possible to interpret $\hat{\beta}^3$ or $\hat{\beta}^4$ as instrumental variable estimators. We will instead refer to $\hat{\beta}^3$ and $\hat{\beta}^4$ as reduced form estimators, relying as they do on the reduced formed probability p_t.

We now consider the conditional expectation correction estimator of Amemiya (1978a) and Heckman (1976b). Assume that η_t in (1) has conditional expectation:

$$E(\eta_t | \delta_t) = g(X_t, \delta_t, p_t)\gamma \tag{15}$$

where g is a differentiable function of δ_t, and the reduced form variables $X_t = (X_{1t}, X_{2t})$ and p_t. For convenience, we write $E(\eta_t | \delta_t)$ in (15) in linear-in-parameters form with γ a column vector of g unknown parameters. Denote $\hat{\eta}_t = \eta_t - E(\eta_t | \delta_t)$. It then follows that $E(\hat{\eta}_t | \delta_t) = 0$. Write equation (1) as:

$$y_t = \delta_t X_{1t}\beta_1 + (1 - \delta_t)X_{2t}\beta_2 + g(X_t, \delta_t, p_t)\gamma + v_t^5 \quad \text{where} \tag{16}$$

$$v_t^5 = \eta_t - g(X_t, \delta_t, p_t)\gamma .$$

When p_t is replaced by its estimate, we obtain:

$$y_t = \delta_t X_{1t}\beta_1 + (1 - \delta_t)X_{2t}\beta_{2t} + g(X_t, \delta_t, \hat{p}_t)\gamma + v_t^6 \quad \text{where} \tag{17}$$

$$v_t^6 = v_t^5 - g(X_t, \delta_t, \hat{p}_t)\gamma$$

$$= \eta_t - g(X_t, \delta_t, p_t)\gamma - [g(X_t, \delta_t, \hat{p}_t) - g(X_t, \delta_t, p_t)]\gamma .$$

Equations (16) and (17) are written in matrix notation as:

$$y = W(\delta)\beta + W_g\gamma + v^5 \quad \text{and} \quad y = W(\delta)\beta + W_{\hat{g}}\gamma + v^6$$

where $\quad W_g = <g(X_t, \delta_t, p_t)> \quad$ and $\quad W_{\hat{g}} = <g(X_t, \delta_t, \hat{p}_t)>$.

Table 5.1 collects the two-stage estimators.

Table 5.1
Two-stage estimators for $y = W(\delta)\beta + \eta$

Instrumental Variables

$$\hat{\beta}^1 - \beta = [W'W(\delta)]^{-1}W'v^1 \qquad v^1 = \eta \tag{1}$$

Instrumental Variables Estimated

$$\hat{\beta}^2 - \beta = [\hat{W}'W(\delta)]^{-1}\hat{W}'v^2 \qquad v^2 = \eta \tag{2}$$

Reduced Form

$$\hat{\beta}^3 - \beta = [W(p)'W(p)]^{-1}W(p)'v^3 \qquad v^3 = \eta + (W(\delta) - W(p))\beta \tag{3}$$

Reduced Form Estimated

$$\hat{\beta}^4 - \beta = [W(\hat{p})'W(\hat{p})]^{-1}W(\hat{p})'v^4 \qquad v^4 = \eta + [W(\delta) - W(p)]\beta - [W(\hat{p}) - W(p)]\beta \tag{4}$$

Amemiya-Heckman[a]

$$\hat{\beta}^5 - \beta = [W(g)'W(g)]^{-1}W(g)'v^5 \qquad v^5 = \hat{\eta} = \eta - E(\eta|\delta) \tag{5}$$

Amemiya-Heckman Estimated

$$\hat{\beta}^6 - \beta = [W(\hat{g})'W(\hat{g})]^{-1}W(\hat{g})'v^6 \qquad v^6 = \hat{\eta} + (W_g - W_{\hat{g}})\gamma \tag{6}$$

---

[a]$W(g) \equiv [W(\delta), W_g], \quad W(\hat{g}) \equiv [W(\delta), W_{\hat{g}}]$.

5.3. Asymptotic distributions

To derive the limiting distribution for each estimator, we assume: (A1) f is differentiable, (A2) β is interior in a compact parameter space, (A3) X_t is uniformly bounded with a convergent empirical distribution function, and (A4) plim $T^{-1}[W(p)'W(p)] = A$, finite and positive definite. We note that plim $T^{-1}[W(\delta)'W(p)] = A$ as $E(\delta_t) = p_t$. It follows that

$$\text{plim } T^{-1}[W'W] = \text{plim } T^{-1}[W'W(\delta)] =$$

$$\text{plim } T^{-1}[W(\delta)'W(p)] \cdot (\text{plim } T^{-1}[W(p)'W(p)])^{-1}$$

$$\cdot \text{plim } T^{-1}[W(p)'W(\delta)] = AA^{-1}A = A .$$

Furthermore, plim $T^{-1}[\hat{W}'\hat{W}] = A$ follows from Lemma 4 of Amemiya (1973) using the consistency of $\hat{p}_t$ for p_t.

For the asymptotic distribution of the Amemiya-Heckman estimator, we assume: (A5) g is differentiable, (A6) γ is interior in a compact parameter space, and (A7) plim $T^{-1}[W(g)'W(g)] = B$, finite and positive definite. From Lemma 4 of Amemiya (1973), we may conclude that plim $T^{-1}[W(\hat{g})'W(\hat{g})] = B$ as well.

We now postulate an error structure for the probability model: (A8) $p_t = Prob[y_t^* < 0] = f[X_t, \alpha]$ where f is a given function of the exogenous variables, X_t, and a column vector of parameters α. If the probability model (A8) is estimated by maximum likelihood, then the following approximation results.

Lemma 5.3.1. Let $\hat{p}_t$ be the estimated value of p_t, i.e., $\hat{p}_t = f[X_t, \hat{\alpha}]$ where $\hat{\alpha}$ is the maximum likelihood estimate of α. Let $\hat{p} = <\hat{p}_t>$, $p = <p_t>$, and $\delta = <\delta_t>$. Then:

$$(\hat{p} - p) \simeq Y'VYD_0^{-1}(\delta - p) \qquad \text{where} \tag{18}$$

$$D_0 = \text{diag}\{p_t(1 - p_t)\}, \quad V = E[(\hat{\alpha} - \alpha)(\hat{\alpha} - \alpha)'] \quad \text{and}$$

$$Y = [\partial p_1/\partial\alpha, \partial p_2/\partial\alpha, ..., \partial p_T/\partial\alpha].$$

Proof. The log-likelihood function, L, is given by:

$$L = \frac{1}{T}\sum_t \delta_t \ln p_t + (1 - \delta_t)\ln(1 - p_t) \tag{19}$$

The gradient is then:

$$L_\alpha = \frac{1}{T}\sum_t \left[(\delta_t/p_t)\,\partial p_t/\partial\alpha - [(1 - \delta_t)/(1 - p_t)]\,\partial p_t/\partial\alpha \right]$$

$$= \frac{1}{T}\sum_t \left[(\delta_t - p_t)/p_t(1 - p_t) \right]\partial p_t/\partial\alpha$$

$$= T^{-1}YD_0^{-1}(\delta - p) \tag{20}$$

A first-order Taylor's expansion of $\hat{p}_t$ around p_t yields $\hat{p} - p_t \simeq (\partial p_t/\partial\alpha')(\hat{\alpha} - \alpha)$. Applying the usual asymptotic argument, we have:

$$\hat{\alpha} - \alpha \overset{D}{=} -L_{\alpha\alpha}^{-1}L_\alpha \qquad \text{so that}$$

$$\hat{p}_t - p_t \overset{D}{=} -(\partial p_t/\partial\alpha')L_{\alpha\alpha}^{-1}T^{-1}YD_0^{-1}(\delta - p) \tag{21}$$

As $V = T^{-1}\hat{L}_{\alpha\alpha'}^{-1}$, we find $\hat{p} - p \simeq Y'VYD_0^{-1}(\delta - p)$ Q.E.D.

Example 5.3.1. The binary logit model is given by:

$$p_t = f[X_t, \alpha] = 1/(1 + e^{-X_t \alpha}).$$

In this case we find $\partial p_t/\partial \alpha = p_t(1 - p_t)X_t$ so that:

$$Y = [\partial p_1/\partial \alpha, \partial p_2/\partial \alpha, ..., \partial p_T/\partial \alpha] = X'D_0$$

where $X = \langle X_t \rangle$. Application of Lemma 5.3.1 yields

$$\hat{p} - p \overset{D}{=} D_0 X V X' D_0 D_0^{-1}(\delta - p) = D_0 X V X'(\delta - p).$$

Throughout the remainder of this chapter, we assume that p_t is given by the binary logit model. To return to the more general case, one need only make the substitution $X = D_0^{-1}Y'$.

The two-stage estimators presented in Section 5.2 are of the form $\hat{\beta}^k - \beta = (W_1^k)^{-1}(W_2^{k'}v^k)$ for appropriate choices of the matrices W_1^k and W_2^k. To derive the asymptotic distribution of $T^{-\frac{1}{2}}(\hat{\beta}^k - \beta)$, we consider the limiting distribution of $T^{-\frac{1}{2}}(W_2^{k'}v^k)$. Assumptions (A1) through (A7) are sufficient for the Lindberg-Feller central limit theorem, which implies:

$$T^{-\frac{1}{2}}(W_2^{k'}v^k) \overset{D}{\to} N[0, \lim E(W_2^{k'}v^k v^{k'}W_2^k)]$$

$$\overset{D}{\to} N[0, \lim E[W_2^{k'} E(v^k v^{k'}|W_2^k) W_2^k]].$$

Lemma 5.3.2 evaluates $E[v^k v^{k'}|W_2^k]$ for v^k given in Table 5.1.

Lemma 5.3.2. Let $\Sigma = E(\eta\eta')$ be a given diagonal matrix. Let:

$$D_1 = \text{diag}\{X_{1t}\beta_1 - X_{2t}\beta_2\}, \qquad D_2 = \text{diag}\{g'(X_t, \delta_t, p_t)\gamma\},$$

$$D_3 = E[\eta(\delta - p)'] = \text{diag}\{E[\eta_t(\delta_t - p_t)]\},$$

$$D_4 = \text{diag}\{E(\eta_t^2|\delta_t)\}, \qquad D_5 = \text{diag}\{[E(\eta_t|\delta_t)]^2\}. \quad \text{Then:}$$

(a) $E(v^1 v^{1'}|X, p) = \Sigma$

(b) $E(v^2 v^{2'}|X, p) = \Sigma$

(c) $E(v^3 v^{3'}|X, p) = \Sigma + 2D_1 D_3 + D_1^2 D_0$

(d) $E(v^4 v^{4'}|X, p) = \Sigma + 2D_1 D_3 + D_1^2 D_0$

$$- [D_1 D_0 X V X' D_3 + D_3 X V X' D_0 D_1 + D_1 D_0 X V X' D_0 D_1]$$

(e) $E(v^5 v^{5\prime}|X, \delta, p) = D_4 - D_5$

(f) $E(v^6 v^{6\prime}|X, \delta, p) = D_4 - D_5 + D_2 D_0 XVX' D_0 D_2$

Proof.

(a) and (b) $v_t^1 = v_t^2 = \eta_t$ so that
$$E(v^1 v^{1\prime}|X, p) = E(v^2 v^{2\prime}|X, p) = \Sigma.$$

(c) $v_t^3 = \eta_t + (\delta_t - p_t)(X_{1t}\beta_1 - X_{2t}\beta_2)$ so that
$$\begin{aligned}
E(v^3 v^{3\prime}|X, p) &= E[\eta + D_1(\delta - p)][\eta' + (\delta - p)'D_1] \\
&= E(\eta\eta') + D_1 E[(\delta - p)\eta'] + E[\eta(\delta - p)']D_1 \\
&\quad + D_1 E[(\delta - p)(\delta - p)']D_1 \\
&= \Sigma + D_1 D_3 + D_3 D_1 + D_1 D_0 D_1 \\
&= \Sigma + 2D_1 D_3 + D_1^2 D_0
\end{aligned}$$
as $E[(\delta_t - p_t)^2] = p_t(1 - p_t)$ and $E[(\delta_t - p_t)(\delta_s - p_s)] = 0$ for $t \neq s$ imply
$E[(\delta - p)(\delta - p)'] = D_0$.

(d) $v_t^4 = \eta_t + (\delta_t - p_t)(X_{1t}\beta_1 - X_{2t}\beta_2) - (\hat{p}_t - p_t)(X_{1t}\beta_1 - X_{2t}\beta_2)$ so that
$$v^4 = \eta + D_1(\delta - p) - D_1(\hat{p} - p).$$

Application of Lemma 5.3.1 and Example 5.3.1 yields
$$\begin{aligned}
v^4 &= \eta + D_1(\delta - p) - D_1 D_0 XVX'(\delta - p) \\
&= \eta + D_1[I - D_0 XVX'](\delta - p). \qquad \text{Then:}
\end{aligned}$$
$$\begin{aligned}
E(v^4 v^{4\prime}|X, p) &= E[\eta + D_1(I - D_0 XVX')(\delta - p)] \cdot \\
&\quad [\eta' + (\delta - p)(I - XVX'D_0)D_1] \\
&= \Sigma + D_1(I - D_0 XVX')D_0(I - XVX'D_0)D_1 \\
&\quad + D_1(I - D_0 XVX')D_3 + D_3(I - XVX'D_0)D_1.
\end{aligned}$$
But $D_1(I - D_0 XVX')D_0(I - XVX'D_0)D_1 = D_1^2 D_0 - D_1 D_0 XVX'D_0 D_1$

since $V = (X'D_0 X)^{-1}$. Furthermore
$$D_1^2 D_0 - D_1 D_0 XVX'D_0 D_1 = D_1 D_0^{\frac{1}{2}} [I - D_0^{\frac{1}{2}} XVX'D_0^{\frac{1}{2}}] D_0^{\frac{1}{2}} D_1.$$

is a positive definite matrix.

(e) $v_t^5 = \hat{\eta}_t = \eta_t - E(\eta_t | \delta_t)$.

Thus $E[\hat{v}_t | X, \delta, p] = 0$ and $E(v^5 v^{5\prime} | X, \delta, p) = D_4 - D_5$.

(f) $v_t^6 = \hat{\eta}_t - [g(X_t, \delta_t, \hat{p}_t) - g(X_t, \delta_t, p_t)]\gamma$.

A first-order Taylor's approximation to $g(X_t, \delta_t, \cdot)$ implies

$$g(X_t, \delta_t, \hat{p}_t) - g(X_t, \delta_t, p_t) \simeq g'(X_t, \delta_t, p_t) (\hat{p}_t - p_t)$$

where $g'(X_t, \delta_t, \cdot) = \partial g(X_t, \delta_t, s)/\partial s$.   Hence:

$$v_t^6 \simeq \hat{\eta}_t - [g'(X_t, \delta_t, p_t)\gamma](\hat{p}_t - p_t).$$

Application of Lemma 5.3.1 and Example 5.3.1 yields

$$v^6 \simeq \hat{\eta} - D_2(\hat{p} - p) = \hat{\eta} - D_2 D_0 XVX'(\delta - p) .$$

As $E[\hat{\eta}(\delta - p)' | X, \delta, p] = 0$, we find:

$$E(v^6 v^{6\prime} | X, \delta, p) = D_4 - D_5 + D_2 D_0 XVX'D_0 XVX'D_0 D_2$$
$$= D_4 - D_5 + D_2 D_0 XVX'D_0 D_2 \qquad \text{Q.E.D.}$$

From Lemma 5.3.2, we obtain the asymptotic distributions for the estimators in Table 5.1.

Theorem 5.3.1. Suppose $\Sigma = E(\eta\eta') = \sigma^2 I$.   Denote

$B_1 = \text{plim } [T^{-1}W(p)'(2D_1 D_3 + D_1^2 D_0)W(p)]$

$B_2 = \text{plim } [T^{-1}W(p)'(D_1 D_0 XVX'D_3 + D_3 XVX'D_0 D_1$

$$+ D_1 D_0 XVX'D_0 D_1)W(p)]$$

$B_3 = \text{plim } [T^{-1}W(g)'D_2 D_0 XVX'D_0 D_2 W(g)]$

$B_4 = \text{plim } [T^{-1}W(g)'D_4 W(g)]$

$B_5 = \text{plim } [T^{-1}W(g)'D_5 W(g)]$.   Then:

(a) $T^{-\frac{1}{2}}(\hat{\beta}^1 - \beta) \xrightarrow{D} N[0, \sigma^2 A^{-1}]$

(b) $T^{-\frac{1}{2}}(\hat{\beta}^2 - \beta) \xrightarrow{D} N[0, \sigma^2 A^{-1}]$

(c) $T^{-\frac{1}{2}}(\hat{\beta}^3 - \beta) \xrightarrow{D} N[0, \sigma^2 A^{-1} + A^{-1}B_1 A^{-1}]$

(d) $T^{-\frac{1}{2}}(\hat{\beta}^4 - \beta) \xrightarrow{D} N[0, \sigma^2 A^{-1} + A^{-1}B_1 A^{-1} - A^{-1}B_2 A^{-1}]$

(e) $T^{-\frac{1}{2}}(\hat{\beta}^5 - \beta) \xrightarrow{D} N[0, B^{-1}[B_4 - B_5]B^{-1}]$

(f) $T^{-\frac{1}{2}}(\hat{\beta}^6 - \beta) \xrightarrow{D} N[0, B^{-1}[B_4 - B_5]B^{-1} + B^{-1}B_3B^1]$

Proof.

(a) plim $(T^{-1}[W'W(\delta)])^{-1} = A^{-1}$. Also

$\quad$ lim $E[T^{-1}W'v^1v^{1'}W] = $ lim $[T^{-1}W'E(v^1v^{1'})W] = \sigma^2 A$. Then:

$\quad T^{-\frac{1}{2}}(\hat{\beta}^1 - \beta) = [T^{-1}W'W(\delta)]^{-1} [T^{-\frac{1}{2}}W'v^1] \xrightarrow{D} N[0, \sigma^2 A^{-1}]$.

In parts (b) to (f), we calculate the appropriate probability limits but omit the details regarding the derivation of the asymptotic distribution.

(b) plim $(T^{-1}[\hat{W}'W(\delta)])^{-1}$ $\qquad = A^{-1}$

$\quad$ plim $E[T^{-1}\hat{W}'v^2v^{2'}\hat{W}]$ $\qquad = \sigma^2 A$

(c) plim $(T^{-1}[W(p)'W(\hat{p})])^{-1}$ $\qquad = A^{-1}$

$\quad$ plim $E[T^{-1}W(p)'v^3v^{3'}W(p)]$ $\qquad = \sigma^2 A + B_1$

(d) plim $(T^{-1}[W(\hat{p})'W(\hat{p})])^{-1}$ $\qquad = A^{-1}$

$\quad$ plim $E[T^{-1}W(\hat{p})'v^4v^{4'}W(\hat{p})]$ $\qquad = \sigma^2 A + B_1 - B_2$

(e) plim $(T^{-1}[W(g)'W(g)])^{-1}$ $\qquad = B^{-1}$

$\quad$ plim $E[T^{-1}W(g)'v^5v^{5'}W(g)]$ $\qquad = B_4 - B_5$

(f) plim $(T^{-1}[W(\hat{g})'W(\hat{g})])^{-1}$ $\qquad = B^{-1}$

$\quad$ plim $E[T^{-1}W(\hat{g})'v^6v^{6'}W(\hat{g})]$ $\qquad = B_4 - B_5 + B_3$ $\quad$ Q.E.D.

It is useful to find the asymptotic distributions of the six estimators under the hypothesis that η and σ are uncorrelated. This is accomplished in Corollary 5.3.1.

Corollary 5.3.1. Define $B_1^N = $ plim $[T^{-1}W(p)'D_1^2D_0W(p)]$, and $B_2^N = $ plim $[T^{-1}W(p)'(D_1D_0XVX'D_0D_1)W(p)]$. Under the null hypothesis in which $E(\eta_t | \delta_t) = 0$

(a) $T^{-\frac{1}{2}}(\hat{\beta}^1 - \beta) \xrightarrow{D} N[0, \sigma^2 A^{-1}]$

(b) $T^{-\frac{1}{2}}(\hat{\beta}^2 - \beta) \xrightarrow{D} N[0, \sigma^2 A^{-1}]$

(c) $T^{-\frac{1}{2}}(\hat{\beta}^3 - \beta) \xrightarrow{D} N[0, \sigma^2 A^{-1} + A^{-1}B_1^N A^{-1}]$

(d) $T^{-\frac{1}{2}}(\hat{\beta}^4 - \beta) \xrightarrow{D} N[0, \sigma^2 A^{-1} + A^{-1}B_1^N A^{-1} - A^{-1}B_2^N A^{-1}]$

(e) $T^{-\frac{1}{2}}(\hat{\beta}^5 - \beta) \xrightarrow{D} N[0, \sigma^2 B^{-1}]$

(f) $T^{-\frac{1}{2}}(\hat{\beta}^6 - \beta) \xrightarrow{D} N[0, \sigma^2 B^{-1}]$

Proof. $E(\eta_t | \delta_t) = 0$ so that $D_3 = D_5 = 0$. As $E(\eta_t | \delta_t) = g(X_t, \delta_t, p_t)\gamma$, it follows that $\gamma = 0$ and hence:

$D_2 = \text{diag}\{ g'(X_t, \delta_t, p_t)\gamma \} = 0.$ Furthermore

$D_4 = \text{diag}\{ E[\eta_t^2 | \delta_t] \} = \text{diag}\{ E[\eta_t^2] \} = \sigma^2 I$ so that

$B_4 = \text{plim} [T^{-1}W(g)'D_4 W(g)] = \sigma^2 B$.

Making the appropriate substitutions, Theorem 5.3.1 yields the desired result. Note that B_1^N is positive definite matrix since $D_1^2 D_0$ is diagonal with positive elements and that

$$B_1^N - B_2^N = \text{plim} \left[T^{-1}W(p)'[D_1^2 D_0 - D_1 D_0 X V X' D_0 D_1]W(p) \right]$$

is positive semidefinite following our remarks given in the proof of Lemma 5.3.2(d).   Q.E.D.

5.4. Efficiency comparisons

We now compare the limiting distributions for the two-stage consistent estimation techniques. We first examine the loss of efficiency incurred through the estimation method itself, i.e., by what degree do we bias standard errors when we ignore that p_t is estimated in the first stage. We then consider the efficiency of the estimator types relative to one another. In each case, we consider the null hypothesis in which δ_t is uncorrelated with v_t and the alternative hypothesis of correlation. The former situation is useful in the testing of regime exogeniety.

5.4.1. Intraestimator orderings

Under the alternative hypothesis in which η_t and δ_t are correlated, Theorem 5.3.1 demonstrates that the instrumental variable estimators $\hat{\beta}^1$ and $\hat{\beta}^2$ are not affected by the estimation of the probability of p_t. The reduced form estimators, $\hat{\beta}^3$ and $\hat{\beta}^4$, have distinct limiting distributions. However, the difference in covariance matrices, $-A^{-1}B_2A^{-1}$, is not definite. The Amemiya-Heckman estimator, $\hat{\beta}^5$, underestimates standard errors relative to its estimated counterpart, $\hat{\beta}^6$.

Under the null hypothesis in which η_t and δ_t are uncorrelated, Corollary 5.1 demonstrates that the instrumental variable estimator, $\hat{\beta}^1$ and $\hat{\beta}^2$, and the Amemiya-Heckman estimators, $\hat{\beta}^5$ and $\hat{\beta}^6$, have identical distributions. On the other hand, the reduced form estimator $\hat{\beta}^3$ overestimates standard errors relative to its estimated counterpart $\hat{\beta}^4$. In this case, $V(\hat{\beta}^4) - V(\hat{\beta}^3) = - A^{-1}B_2^N A^{-1} \leqslant 0$ as B_2^N is a positive semidefinite matrix.

5.4.2. Interestimator orderings

Under the alternative hypothesis in which η_t and δ_t are correlated, Theorem 5.3.1 does not permit estimator efficiency orderings between the different estimator types. The difference between the reduced form and instrumental variable estimated covariance matrices, $V(\hat{\beta}^4) - V(\hat{\beta}^3) = A^{-1}(B_1 - B_2)A^{-1}$, is indefinite and similarly the difference between Amemiya-Heckman and instrumental variables estimators is indefinite.

Under the null hypothesis in which η_t and δ_t are uncorrelated, we find that the instrumental variable estimator $\hat{\beta}^2$ (or $\hat{\beta}^1$) dominates the reduced formed estimator $\hat{\beta}^4$ (or $\hat{\beta}^3$) in the sense that $V(\hat{\beta}^4) - V(\hat{\beta}^2) = A^{-1}[B_1^N - B_2^N]A^{-1}$ is a positive definite matrix. It is also easy to see that $V(\hat{\beta}^3) - V(\hat{\beta}^1) = A^{-1}B_1^N A^{-1}$ is positive definite.

Finally, we compare the Amemiya-Heckman and instrumental variable estimators. In this case the parameters of interest are β_1 and β_2 and therefore, we compare the leading $(K_1 + K_2)$ submatrix of $V(\hat{\beta}^6) = \sigma^2 B^{-1}$ with the $(K_1 + K_2)$ matrix $V(\hat{\beta}^2) = \sigma^2 A^{-1}$. Let:

$$B = \begin{bmatrix} C_1 & C_2 \\ C_2' & C_3 \end{bmatrix} \quad \text{with} \quad C_1 = \begin{bmatrix} X_1'PX_1 & 0 \\ 0 & X_2'(I-P)X_2 \end{bmatrix},$$

$C_2 = \text{plim } [T^{-1}X_1'PW_g]$, and $C_3 = \text{plim } [T^{-1}W_g'W_g]$

where $P = \text{diag } \{ p_t \}$. Now:

$$A = \text{plim } T^{-1}[W(p)'W(p)] = \begin{bmatrix} X_1'P^2X_1 & X_1'P(I-P)X_2 \\ X_2'P(I-P)X_1 & X_2'(I-P)^2X_2 \end{bmatrix}$$

so that $\quad C_1 - A = [X_1, - X_2]'[P(I - P)][X_1, - X_2] \geqslant 0$.

The Amemiya-Heckman estimated covariance matrix for (β_1, β_2) exceeds the estimated instrumental variable when $B^{11} - A^{-1} \geqslant 0$. As

$B^{11} = [C_1 - C_2 C_3^{-1} C_2']^{-1}$,   we have the equivalent condition

$A \geqslant C_1 - C_2 C_3^{-1} C_2'$ or $C_2 C_3^{-1} C_2' \geqslant C_1 - A \geqslant 0$.

In conclusion, the instrumental variable estimator will dominate the Amemiya-Heckman estimator provided the loss in efficiency incurred in using an instrumental variable estimator relative to the least squares estimator is smaller than the loss in efficiency incurred through the addition of an irrelevant regressor.

CHAPTER 6

Estimation of the demand for electricity and natural gas from billing data

6.1. Introduction

In this chapter we estimate a residential demand model for electricity and natural gas using data from the National Interim Energy Consumption Survey conducted during the year 1978-1979. This study differs from previous work in two major respects. First, we utilize individual monthly billing data in a pooled time-series cross-section framework. This permits a seasonally disaggregated analysis that considers individual behavior over time. Second, we use the engineering thermal load of Chapter 2 to model the household production technology for space heating and air conditioning, allowing more precise separation of the effects of economic variables from those of climate.

The thermal load technique is combined with billing cycle data in our study in two ways. First, we use the thermal model to estimate billing cycle load on a household-by-household basis. In this approach, two households with equivalent building characteristics facing identical weather patterns would be predicted to have the same energy demand. In this way we adopt a strategy of incorporating an engineering thermal projection into our energy demand analysis. In reality, we realize that the demands may vary significantly between otherwise identical households due to differences in income, household size, activity patterns, and the cost of energy. Thus departures from the engineering estimates are due to socioeconomic sensitivity in the rate of appliance stock utilization.

Secondly, we use the engineering thermal load techniques to estimate the cost of comfort. Here the estimated change in energy input required to effect a one-degree change in ambient temperature is multiplied by the marginal price of the fuel input. In the next section we combine the engineering economic approach in an econometrically estimable model.

Section 3 presents the electricity demand and natural gas estimation while Section 4 concludes with an analysis of energy demand under the ASHRAE conservation standards discussed in Chapter 3. Detailed descriptions of the data preparation procedures are given in Appendix A.

6.2. Speciation of conditional demand models

An econometric conditional demand model is developed by noting that a household's total electricity consumption in any period is simply the sum of the electricity used by each appliance in that period:

$$Y_{it}^f = \sum_{j=1}^{J} UEC_{it}^j \, \delta_i^j \, (X_{it}^j \beta^j) + Z_i \gamma + \varepsilon_{it} \tag{1}$$

where Y_{it}^f is the demand for fuel in period t by household i, UEC_{it}^j is the unit energy consumption of appliance j in period t by household i, δ_i^j is the indicator of appliance j ownership by household i, X_{it}^j is a vector of socioeconomic variables affecting utilization of appliance j by household i in period t, β is a vector of parameters associated with X_{it}^j, Z_i is a vector of socioeconomic variables affecting time-independent usage of fuel, γ is a vector of parameters associated with Z_i, and ε_{it} is the error term for household i in period t.

The term $Z_i \gamma$ accounts for the presence of electric refrigerators, ovens, ranges, microwave ovens, freezers, washers, and clothes dryers. For our purposes, the UEC's associated with these appliances are of secondary interest only and we view β as the parameters of interest.

A pure conditional demand approach approximates the terms UEC_{it}^j by functions of variables related to the technology of the appliance. A common specification for the UEC of space conditioning represents this term as a linear function in square feet, insulation levels, heating degree-days, etc. To illustrate this approach we write:

$$UEC_{it}^j = H_{it}^j \alpha^j + v_{it}^j \tag{2}$$

where H_{it}^j is a vector of characteristics of appliance j for household i in period t, α^j is a vector of parameters associated with H_{it}^j, and v_{it}^j is the error term in linear specification of UEC. Equations (1) and (2) imply:

$$Y_{it}^f = \sum_{j=1}^{J} \delta_i^j \, (H_{it}^j \alpha^j) \, (X_{it}^j \beta^j) + Z_i \gamma + \varepsilon_{it} + \tilde{v}_{it} \qquad \text{where} \tag{3}$$

$$\tilde{v}_{it} = \sum_{j=1}^{J} \delta_i^j \, v_{it}^j \, (X_{it}^j \beta^j)$$

The purpose of the engineering econometric approach is to minimize the effect of the measurement error $\tilde{v}_{it}$ through a thorough thermal modeling of the space conditioning appliance technology. We argue that the engineering econometric approach is superior to the pure conditional demand methodology because it efficiently and effectively incorporates all relevant available engineering data and emphasizes the structure of the estimated equation.

Our empirical work focuses on space and water heating and air conditioning as the end-uses that are the largest contributors to cross-sectional and seasonal variation in energy consumption. To reduce the number of parameters to be estimated, we subtract the term $Z_i\gamma$ from the left-hand side of equation (1), using values of γ reported in the literature. If the UEC values γ are measured with error, this approach may reduce the precision of the estimates of the parameters, β^j, but will not bias their estimates if measurement errors are uncorrelated with the explanatory variables. The form of equation (1) used in our study is:

$$Y_{it}^f - Z_i\gamma = UEC_{it}^{SH} \delta_i^{SH} (X_{it}^{SH}\beta^{SH}) + UEC_{it}^{AC} \delta_i^{AC} (X_{it}^{AC}\beta^{AC})$$
$$+ UEC_{it}^{WH} \delta_i^{WH} (X_{it}^{WH}\beta^{WH}) + X_{it}^M\beta^M + \varepsilon_{it} \qquad (4)$$

where the superscripts *SH*, *AC*, *WH*, and *M* denote space heating, air conditioning, water heating, and miscellaneous uses respectively. The "miscellaneous" term captures consumer response for all end-uses, other than space and water heating and air conditioning, that are not contained in Z_i.

Several features of our empirical specification deserve note. First, the expression $UEC_{it}^{SH}\delta_i^{SH}$ is in reality an inner product of several space-heating system with a corresponding vector of dummy indicator variables that select the chosen alternative. This observation is important for space-heating systems as several system types are capable of providing the same delivered service. We do not, however, differentiate the utilization factor for space heating by system type. This is due primarily to computational considerations but also is implied by our theoretical structure, which incorporates engineering attributes of alternative systems in the calculation of UEC_{it}^{SH}. For air conditioning and water heating, we follow the model of Chapter 3 and consider one system type for each fuel type.

Our second observation concerns the specification of the error ε_{it}. We write equation (4) as

$$Y_{it}^f - Z_i\gamma = Y_{it}^{SH} + Y_{it}^{AC} + Y_{it}^{WH} + Y_{it}^M \qquad \text{where} \qquad (5)$$

$$Y_{it}^{SH} = UEC_{it}^{SH} \delta_i^{SH} (X_{it}^{SH}\beta^{SH}) + \varepsilon_{it}^{SH}$$

$$Y_{it}^{AC} = UEC_{it}^{AC} \delta_i^{AC} (X_{it}^{AC}\beta^{AC}) + \varepsilon_{it}^{AC}$$

$$Y_{it}^{WH} = UEC_{it}^{WH} \delta_i^{WH} (X_{it}^{WH}\beta^{WH}) + \varepsilon_{it}^{WH} \qquad \text{and}$$

$$Y_{it}^M = X_{it}^M \beta^M + \varepsilon_{it}^M$$

denote respectively the fuel demands for space heating, air conditioning, water heating, and other uses. The error ε_{it} is precisely $\varepsilon_{it}^{SH} + \varepsilon_{it}^{AC} + \varepsilon_{it}^{WH} + \varepsilon_{it}^M$. This representation allows correlation of each component error term for fuel usage with the random components of utility that affect the presence or absence of a given HVAC system. In practice some correlation of unobserved variables is likely. For an appliance such as an air conditioner, an unobserved effect that increases the utility of the service supplied by the appliance (e.g., poor natural ventilation in the dwelling) is likely to increase both its probability of selection and its intensity of use. For an appliance such as a water heater, unobserved factors that increase intensity of use (e.g., tastes for hot-water clothes washing) are likely to decrease the probability of choosing the electric alternative that has a higher operating-to-capital cost ratio than the alternative fuel.

Dubin and McFadden (1984) introduced a family of models that permit consistent estimation of the parameters in equation (1). Their specification assumed that the choice model was multivariate logistic. Appendix B extends their analysis to include the generalized extreme value family of disturbances. The specification of fuel demand given in equation (5) is then consistent with the nested logistic choice models employed in Chapter 3.

We assume that each component error term for fuel usage has a conditional expectation that is linear in the choice model unobservables and has a conditional variance that is constant. Following the discussion in Chapter 5, we adopt the Amemiya-Heckman selectivity correction method and include estimated conditional expectations of the equation error given the observed portfolio choice as additional explanatory variables. Appendix B demonstrates that each conditional expectation is a simple function in the probabilities estimated from the nested logit model.

Our stochastic specification further assumes that each component fuel demand (i.e., fuel demand for space heating, fuel demand for air conditioning, etc.), follows a binary switching regression. The first regime represents the observed choice and the alternate regime represents collectively the unchosen alternatives. In the electricity demand equation, for example, δ_i^{WH} denotes the presence of an electric water-heating system. Given the structure in Chapter 3, electric water heating is selected if and only if its utility exceeds the utility of gas or oil systems. In the electricity demand equation, we treat gas and oil water-heating systems simply as "non electric." The conditional logistic structure of water-heating fuel choice given space-heating system choice permits the aggregation of unchosen alternatives into a generic category whose deterministic utility is the exponentially weighted average of the primary utilities. In estimation we therefore include three conditional expectation terms, one each for space heating, water heating, and air conditioning, as additional explanatory variables. We assume that the component error term, ε_{it}^{M}, is uncorrelated with the choice of HVAC systems.

The time-series cross-section structure of the billing data permits separate estimates of the correlation of fuel usage and appliance choice by season. These estimates refine the empirical evidence obtained in Dubin and McFadden (1984), which relied on annual data. The separate estimation of electricity and natural gas allows the appliance endogeniety issue to be additionally analyzed by fuel type.[1]

6.3. Estimation of the demand for electricity and natural gas

6.3.1. Variable definitions

The form of equation (4) used in estimation contains the unit-energy-consumption terms, UEC_{it}^{j}, as well as the utilization variables represented by $X_{it}^{j}\delta_{i}^{j}$ on the right-hand side, and net consumption, Y_{it}^{f} - $Z_{i}\gamma$, on the left-hand side. In the electricity demand models, the dependent variable, NETQUAN, is calculated as the difference QUAN minus QEBASE. Using typical UEC's values, we assume:

QEBASE (kwh) = [3.99 (number of automatic refrigerators) + 1.87 (number of manual refrigerators) + 2.8 (number of ovens) + 0.5 (microwave oven) + 3.8 (separate food freezer) + 0.9 (dishwasher) + 0.2 (clothes washer)] × [number of days in billing period]   and

---

[1]We recognize that the time-series cross-section structure of the billing data permits a stochastic specification in which individual effects induce correlation over time. Viewing individual demand equations as Seemingly Unrelated Regressions (SUR) should increase efficiency in the estimation. In this study, we estimate the demand equations as separate cross sections, focusing on the differences in parameter estimates over the billing cycle rather than on the possibilities for pooling.

QGBASE (therms) = [0.114 (clothes dryer) + 0.319 (first gas oven) + 0.160 (second gas oven)] × [number of days in billing period].

On the right-hand side, space-heating usage (UEC_{it}^{SH}) is the fuel usage (kwh for electricity, therms for natural gas) required to maintain an indoor temperature of 70 °F. This quantity is calculated by the integration of daily thermal load over the billing cycle and then adjusted for the efficiency of the heating system and for delivery loss. Details are given in Appendix A. Air-conditioning usage (UEC_{it}^{AC}) is the predicted kwh required to maintain the dwelling at 75 °F.

In the specification of the utilization factor for HVAC systems, we use a constant, income, and the marginal price of comfort. The marginal price of comfort represents the cost of increasing indoor temperature by one degree Fahrenheit (decreasing temperature in the case of air conditioning). It is calculated by multiplying the tail-block marginal fuel price by the energy usage required to increase indoor temperature from 69 to 70 degrees.

Water-heating usage (UEC_{it}^{WH}) is calculated using the formula developed in Chapter 3 to calculate operating costs. An adjustment is made to reflect the length of the billing period. The utilization factor for water heating uses the tail-block marginal fuel price, income, and a constant term.

Variables in the "miscellaneous usage" term include an intercept, the number of household members, income, the marginal fuel price, and the number of days in the billing period. Definitions of all variables and constructions used in this analysis are presented in Table 6.1 and 6.2. for electricity and natural gas respectively.

The data used to estimate this model are a subset of the 911 households used in estimating the nested logit model in Chapter 3. Of the 911 households, 862 had some data on electricity demand while 504 had data on natural gas demand. We define the billing periods using information about the starting dates of actual bills. A household is assumed to have an observation in a given calendar month if it has a billing period with a start date in that month. This procedure serves to define relatively precise calendar effects and separates observations into homogeneous groups.

Variable means over billing periods are presented in Tables 6.3 and 6.4 for electricity and natural gas respectively. Results in these and later tables are presented in three groups: "Winter," which contains the months November, December, January, and February; "Summer," which

Table 6.1

Variable definitions (electricity models)

SUSHE	Space-heating energy consumption from thermal model for HVAC systems 13, 14, 15, 18 — electric forced-air without central air conditioning, electric forced-air with central air conditioning, electric heat pump, electric baseboard heat without central air conditioning (kwh)
SUCAC	Air-conditioning energy consumption from thermal model for HVAC systems 2, 8, 14, 15 — gas forced-air with central air, oil forced-air with central air, electric forced-air with central air, electric heat pump (kwh)
SUWHE	Water-heating energy consumption from average usage relationship (see text) for electric hot-water systems (kwh)
DSHE	Space-heating energy required to increase indoor temperature one degree from thermal model (kwh)
DCAC	Air conditioning energy required to decrease indoor temperature one degree from thermal model (kwh)
SHP	Marginal cost of increasing indoor temperature one degree F $\equiv$ DSHE $\times$ MPE ($)
ACP	Marginal cost of decreasing indoor temperature one degree F $\equiv$ DCAC $\times$ MPE ($)
MPE	Marginal cost of electricity ($/kwh)
INCOME	Total annual income (thousands $)
DAYS	Number of days in billing period
ONE	Constant term
NHSLDMEM	Number of household members

contains the months of June, July, August, and September; and "Off-Season," which contains the months of March, April, May, and October.

In reading Tables 6.3 and 6.4 one should observe that the mean values for the unit energy consumption variables and their interactions reflect the indicator for HVAC system type. Thus the variables SUSHE, SUSHEP, and SUSHEY should be divided by the proportion of households with electric space heating, SHE, in order to determine the mean values for the subpopulation consisting only of households with electric space heating.

Generally, the SUSHE variables increase in the colder months and decrease in the warmer months, as expected. The SUCAC variables reverse this pattern, exhibiting their largest magnitudes during the summer months. The variables DSHE and DCAC represent the energy usage (in kwh and therms respectively) required to change indoor temperature by one degree. SHP and ACP are the costs of making these changes.

Table 6.1—continued

SUSHEP	SUSHE × SHP
SUSHEY	SUSHE × INCOME
SUCACP	SUCAC × ACP
SUCACY	SUCAC × INCOME
CAC	Central air-conditioning dummy
WHE	Water-heating electric dummy
SHE	Space-heating electric dummy
SUWHE	SUWHE × MPE
SUWHEP	SUWHE × INCOME
H1	[(PSHE-SHE) · log (PSHE)/PSHNE - (PSHNE - SHNE) · log (PSHNE)/PSHE] where PSHE is the estimated probability of choosing electric space heating, PSHNE is the estimated probability of choosing non-electric space heating, SHE is an alternative specific variable for electric space heating, and SHNE = 1 - SHE.
H2	[(PCAC - CAC) · log (PCAC)/PNCAC - (PNCAC - NCAC) · log (PNCAC/PCAC] where PCAC is the estimated probability of choosing central air conditioning, PNCAC is the estimated probability of not choosing central air conditioning, CAC is an alternative specific variable for central air conditioning, and NCAC = 1 - CAC.
H3	[(PWHE-WHE) · log (PWHE)/PWHNE - (PWHNE - WHNE) · log (PWHNE)/PWHE] where PWHE is the estimated probability of choosing electric water heating, PWHNE is the estimated probability of choosing non-electric water heating, WHE is an alternative specific variable for electric water heating, and WHNE = 1 - WHE.
NETQUAN	Difference between actual consumption, QUAN, and base consumption, QEBASE (kwh)

Interestingly, SHP varies from a high value of $1.27 for the month of November to essentially zero in the summer months. The peak cost of air-conditioning services occurs in July at $1.15 per degree per billing period. Inspection of Table 6.4 indicates that the marginal cost for heating services is at its highest in December at the level of $2.48.

6.4. Estimation results

Ordinary least squares estimates of the demand models by billing period are presented in Tables 6.5 and 6.6 for electricity and natural gas respectively. Interpretation of coefficients on miscellaneous usage variables is straightforward. The number of household members, NHSLDMEM,

Table 6.2

Variable definitions (natural gas models)

SUSHG	Space-heating energy consumption from thermal model for HVAC systems 1, 2, 3 — gas forced-air with central air conditioning, gas forced-air without central air conditioning, gas hot water without central air conditioning (therms)
SUWHG	Water-heating energy consumption from usage relationship (see text) for gas hot water systems (therms)
DSHG	Energy required to increase indoor temperature one degree from thermal model (therms)
SHP	Marginal cost of increasing indoor temperature one degree F $\equiv$ DSHG $\times$ MPG ($)
MPG	Marginal cost natural gas ($/therm)
INCOME	Total annual income (thousands $)
DAYS	Number of days in billing period
ONE	Constant term
NHSLDMEM	Number of household members
SHG	Space-heating gas dummy
WHG	Water-heating gas dummy
SUSHGP	SUSHG $\times$ SHP
SUSHGP	SUSHG $\times$ INCOME
SUWHGP	SUWHG $\times$ MPG
SUWHGY	SUWHG $\times$ INCOME
H1	[(PSHG-SHG) $\cdot$ log (PSHG)/PSHNG - (PSHNG - SHNG) $\cdot$ log (PSHNG)/PSHG] where PSHG is the estimated probability of choosing gas space heating, PSHNG is the estimated probability of choosing non-gas space heating, SHG is an alternative specific variable for gas space heating, and SHNG = 1 - SHG.
H2	[(PWHG-WHG) $\cdot$ log (PWHG)/PWHNG - (PWHNG - WHNG) $\cdot$ log (PWHNG)/PWHG] where PWHG is the estimated probability of choosing gas water heating, PWHNG is the estimated probability of choosing non-gas water heating, WHG is an alternative specific variable for gas water heating, and WHNG = 1 - WHG.
NETQUAN	Difference between actual consumption, QUAN, and base consumption, QGBASE (therms)

affects usage at the approximate rate of 80 kwh per person per billing period. An additional day in the billing period represents on average an extra 10 kwh not accounted in other effects.[2]

[2]In Tables 6.5 through 6.6, t-statistics for coefficient significance are given in the line following the parameter estimates.

Table 6.3a
Variable means by billing period—Winter

Variable	November	December	January	February
SUSHE	582.6	1118.	1048.	843.1
SUSHEP	3423.	8532.	6406.	4934.
SUSHEY	14870.	29170.	25960.	21870.
SUCAC	9.620	1.392	0.7268	2.603
SUCACP	24.09	1.877	0.2004	1.952
SUCACY	229.4	34.44	13.72	55.68
SUWHE	148.6	165.1	157.5	157.6
SUWHEP	5.497	5.770	5.748	5.446
SUWHEY	3532.	3910.	3736.	3777.
SHP	1.056	1.269	1.259	1.204
ACP	0.0983	0.0165	0.0076	0.02736
DSHE	29.55	39.38	35.63	37.24
DCAC	2.854	0.4594	0.1855	0.7170
MPE	0.0392	0.0392	0.0389	0.0387
INCOME	22.32	22.88	22.38	22.66
NHSLDMEM	3.195	3.238	3.231	3.202
DAYS	34.65	40.21	35.20	34.33
H1	0.03236	0.02728	-0.00964	0.06660
H2	0.06575	-0.03021	0.04316	0.01854
H3	0.06225	0.04874	0.03790	0.07709
SHE	0.2378	0.2349	0.2385	0.2526
WHE	0.4269	0.4108	0.4354	0.4324
CAC	0.4077	0.3570	0.3954	0.3744
QUAN	1200.	1686.	1628.	1527.
QEBASE	465.8	529.6	471.1	463.9
Observations	677	762	650	673

Interpretation of the space-heat interaction terms is less straightforward because they are multiplied by the unit energy consumptions for heating. If the thermal model were to predict exactly the space-heating consumption necessary to maintain an indoor temperature of 70 degrees, and if there were no response to economic variables, then the coefficient on SUSHE would be unity and the other coefficients would be zero. The coefficients on SUSHEP and SUSHEY can be interpreted as the amounts by which changes in the price of comfort and in income respectively modify the effect of unit energy consumption on net usage. The signs on these coefficients conform to the predictions of our model in that price effects are generally negative and income effects are generally positive.

The estimated coefficient of SUSHE varies across the 12 billing periods. However, the magnitude of the average effect suggests that the thermal model may be overestimating usage by more than can be

Table 6.3b
Variable means by billing period—Summer

Variable	June	July	August	September
SUSHE	37.72	7.455	83.93	136.8
SUSHEP	111.8	6.461	395.8	533.0
SUSHEY	935.7	193.6	2074.	3531.
SUCAC	152.9	155.0	163.6	110.7
SUCACP	625.0	944.5	836.8	328.2
SUCACY	4359.	4477.	4752.	2795.
SUWHE	150.7	147.9	159.5	141.6
SUWHEP	5.210	5.495	5.559	5.130
SUWHEY	3541.	3530.	3848.	3378.
SHP	0.1351	0.04286	0.1879	0.4797
ACP	0.9367	1.151	1.094	0.6438
DSHE	7.083	1.882	9.488	14.49
DCAC	57.82	67.11	56.42	24.98
MPE	0.03909	0.03958	0.03873	0.03918
INCOME	22.97	22.73	23.26	22.57
NHSLDMEM	3.251	3.208	3.233	3.187
DAYS	36.32	35.73	36.75	33.30
H1	0.04006	0.002390	0.02343	0.006142
H2	0.01705	0.05941	0.01728	0.08123
H3	0.02807	0.03432	0.04476	0.03720
SHE	0.2347	0.2245	0.2398	0.2332
WHE	0.4025	0.4106	0.4210	0.4198
CAC	0.3711	0.3988	0.3760	0.4052
QUAN	1272.	1352.	2713.	1057.
QEBASE	483.1	471.6	493.5	442.6
Observations	733	677	734	686

accounted for by the effects of price and income. While the significance of individual coefficients varies somewhat in the different periods, we see that the models predict well in the winter and off-season periods but not as well during the summer months. The poor performance of the electricity demand model in the summer is apparently caused by colinearity among the explanatory variables.

A very similar pattern is exhibited in the natural gas estimation. Here the sample is selected to include all households for which gas is an available fuel. An additional household member represents a modest increase of approximately three therms during the billing period. An additional day in the billing period increases consumption by fewer than four therms. The individual coefficients are generally significant but reveal some discrepancies in the expected signs. The general magnitudes of the coefficients suggest again that the thermal model may overestimate utilization, but not in a fashion that prevents its usefulness in forecasting.

Table 6.3c
Variable means by billing period—Off-Season

Variable	March	April	May	October
SUSHE	396.0	273.3	55.10	419.7
SUSHEP	1980.	1174.	141.5	2079.
SUSHEY	10680.	6928.	1402.	10520.
SUCAC	15.24	47.73	97.02	13.47
SUCACP	47.98	163.2	346.4	44.72
SUCACY	305.4	1127.	2460.	314.6
SUWHE	135.1	158.1	145.2	150.0
SUWHEP	4.814	5.486	5.240	5.212
SUWHEY	3323.	3741.	3474.	3596.
SHP	0.8396	0.6755	0.2697	0.8140
ACP	0.1569	0.3680	0.7052	0.1678
DSHE	25.34	23.77	8.825	27.12
DCAC	4.240	13.86	35.44	5.085
MPE	0.03842	0.03926	0.03944	0.03922
INCOME	23.93	23.25	22.25	23.27
NHSLDMEM	3.200	3.257	3.206	3.231
DAYS	31.22	35.77	35.29	35.79
H1	-0.007479	0.02737	0.02305	0.007580
H2	0.1280	0.01261	0.07868	-0.001253
H3	-0.007844	0.08786	0.01397	0.03617
SHE	0.2198	0.2386	0.2308	0.2276
WHE	0.3802	0.4199	0.4031	0.4092
CAC	0.4286	0.3922	0.4046	0.3696
QUAN	1158.	1087.	1039.	1047.
QEBASE	431.5	481.1	464.6	478.1
Observations	455	612	650	782

Regarding the correlation of appliance choice and utilization, we see that the selectivity correction terms vary in significance by billing period and system type. In the electricity demand models, the winter months are associated with positive and significant selectivity coefficients for the endogeniety of space heating. Air-conditioning choice effects are present in the summer months of June and July. Choice of water-heating fuel is indicated to be endogenous in five of 12 months (January, March, May, July, and December). Only two months reveal no correlation between usage and system choice behaviors: April and September.[3]

In the natural gas estimation, space-heating endogeniety is indicated in the months of August and October while water-heating endogeniety is indicated in January, April, July, August, November, and December.

[3]Under the null hypothesis of exogeneity of appliance choice decisions, the usual t-tests need not be corrected for the two-stage estimation method employed.

Table 6.4a
Variable means by billing period—Winter

Variable	November	December	January	February
SUSHG	298.4	375.3	350.6	268.2
SUSHGP	1462.	1279.	1033.	763.6
SUSHGY	8515.	9001.	8349.	6768.
SUWHG	20.95	21.39	18.95	18.32
SUWHGP	4.766	4.742	4.392	4.098
SUWHGY	536.3	503.3	443.2	446.0
SHP	2.456	2.483	2.279	2.098
DSHG	10.86	11.28	10.09	9.520
WHG	0.9260	0.9211	0.9300	0.9246
SHG	0.9644	0.9690	0.9621	0.9623
QGBASE	13.81	13.41	12.75	10.78
QUAN	261.8	340.1	307.2	241.3
MPG	0.2263	0.2223	0.2261	0.2231
INCOME	23.10	22.26	22.34	22.91
NHSLDMEM	3.173	3.254	3.120	3.304
DAYS	40.98	42.19	38.05	35.28
H1	0.4265	0.4150	0.4118	0.4029
H3	0.3507	0.3214	0.3521	0.3439
Observations	365	355	343	345

Table 6.4b
Variable means by billing period—Summer

Variable	June	July	August	September
SUSHG	14.72	22.35	28.66	69.12
SUSHGP	30.96	144.5	108.6	235.4
SUSHGY	378.5	652.5	756.1	1871.
SUWHG	21.23	21.40	20.04	18.49
SUWHGP	4.829	4.929	4.524	4.226
SUWHGY	512.6	538.9	486.8	452.0
SHP	0.7331	0.6699	0.9198	1.377
DSHG	3.241	2.972	3.920	5.890
WHG	0.9321	0.9246	0.9263	0.9278
SHG	0.9674	0.9623	0.9632	0.9639
QGBASE	14.13	13.15	13.12	11.93
QUAN	54.00	58.63	59.24	75.66
MPG	0.2248	0.2247	0.2217	0.2242
INCOME	22.85	23.29	22.56	22.98
NHSLDMEM	3.204	3.255	3.224	3.219
DAYS	41.19	42.04	39.15	36.48
H1	0.4191	0.4189	0.4056	0.4121
H3	0.3708	0.3532	0.3412	0.3474
Observations	368	345	353	360

Table 6.4c
Variable means by billing period—Off-Season

Variable	March	April	May	October
SUSHG	146.6	94.46	36.59	155.1
SUSHGP	267.1	230.5	112.2	519.0
SUSHGY	3700.	2397.	945.7	3838.
SUWHG	15.69	18.96	20.92	19.97
SUWHGP	3.138	4.312	4.750	4.461
SUWHGY	384.8	460.8	506.2	480.5
SHP	1.491	1.618	1.114	1.905
DSHG	7.477	7.294	5.000	8.588
WHG	0.9276	0.9242	0.9370	0.9262
SHG	0.9864	0.9606	0.9685	0.9644
QGBASE	8.757	12.65	14.46	12.88
QUAN	159.4	105.3	69.06	153.3
MPG	0.2012	0.2223	0.2239	0.2213
INCOME	23.15	22.88	22.66	22.88
NHSLDMEM	3.063	3.203	3.189	3.209
DAYS	31.24	38.18	41.73	39.16
H1	0.4007	0.4058	0.4345	0.4061
H3	0.2625	0.3326	0.3807	0.3413
Observations	221	330	349	393

These results support the findings of Dubin and McFadden (1984) and indicate that behavioral correlations in HVAC-system choice and utilization occur in those periods in which systems are most intensely used.

To gauge the effects of price and income and to determine the sources of their sensitivity, we calculate short-run price and income elasticities. Tables 6.7 and 6.8 present the elasticities for the electricity and natural gas models respectively. Recall that the estimated model has the form:

$$\text{NETQUAN} = \text{QUAN} - \text{QEBASE} = \text{SUSHE}\gamma_0 + \text{SUSHEP}\gamma_1 + \text{SUSHEY}\gamma_2 + \text{SUCAC}\gamma_3 + \text{SUCACP}\gamma_4 + \text{SUCACY}\gamma_5 + \text{SUWHE}\gamma_6 + \text{SUWHEP}\gamma_7 + \text{SUWHEY}\gamma_8 + \gamma_9 + \text{MPE}\gamma_{10} + \text{INCOME}\gamma_{11} + \text{NHSLDMEM}\gamma_{12} + \text{DAYS}\gamma_{13} + \text{H1}\gamma_{14} + \text{H2}\gamma_{15} + \text{H3}\gamma_{16}.$$

It then follows that the elasticity of total energy usage with respect to price is

$$(\text{MPE/QUAN}) (\partial\text{QUAN}/\partial\text{MPE}) = (\text{MPE/QUAN}) \times$$
$$[(\text{SUSHE})(\text{DSHE}) \gamma_1 + (\text{SUCAC})(\text{DCAC}) \gamma_4 + (\text{SUWHE}) \gamma_7 + \gamma_{10}].$$

Table 6.5a
Electricity demand model—Winter

Variable	November	December	January	February
SUSHE	0.5602	0.2385	0.4380	0.6304
	6.945	5.340	5.996	8.750
SUSHEP	0.004739	0.009684	-0.006719	-0.002097
	0.7972	5.351	-1.159	-4.396
SUSHEY	-0.006487	-0.0002629	-0.003334	0.8217
	-4.304	-0.2433	-2.091	0.5177
SUCAC	-2.283	-5.158	-36.52	3.079
	-1.397	-0.5151	-1.728	0.5773
SUCACP	-0.01747	-0.3552	34.05	3.447
	-0.08841	-0.09030	0.4769	0.9342
SUCACY	0.09381	0.1136	0.3962	-0.3470
	1.446	0.2877	0.4029	-1.265
SUWHE	1.509	4.501	5.148	5.368
	1.761	6.246	4.445	7.164
SUWHEP	-15.81	-75.10	-56.29	-101.5
	-0.8046	-5.029	-2.259	-6.398
SUWHEY	0.009234	0.01835	0.007412	0.3478
	0.7024	1.226	0.3933	2.054
ONE	-219.8	-100.1	-257.8	-507.1
	-1.196	-0.4501	-1.030	-2.307
MPE	-7041.	-5931.	-9895.	6155.
	-1.647	-1.117	-1.698	1.215
INCOME	3.175	4.022	9.962	5.031
	1.113	1.064	2.393	1.422
NHSLDMEM	87.75	79.04	44.83	26.26
	5.006	3.278	1.839	1.181
DAYS	13.27	10.50	17.34	7.984
	6.382	4.645	5.468	2.681
H1	26.08	176.9	306.6	41.48
	0.5790	3.122	4.285	0.7988
H2	39.49	-14.13	27.22	-1.521
	2.054	-0.5057	1.001	-0.06066
H3	3.796	-118.7	-195.7	-122.1
	0.1092	-2.355	-3.819	-2.679
R^2	0.6159	0.7385	0.6956	0.7780
Observations	677	762	650	673
Standard error	642.8	943.6	889.1	808.0

Table 6.5b
Electricity demand model—Summer

Variable	June	July	August	September
SUSHE	-0.2217	1.006	-9.408	0.001985
	-0.4188	0.2847	-0.7851	0.01340
SUSHEP	-0.07204	-0.2988	0.02488	0.02148
	-1.015	-0.09922	0.03204	1.421
SUSHEY	0.03627	-0.05019	0.2226	0.005645
	2.811	-1.236	0.6690	1.368
SUCAC	0.2222	0.6895	-2.066	1.001
	1.126	3.187	-0.2097	4.380
SUCACP	0.1602	0.06957	0.1765	-0.1803
	4.884	3.783	0.1757	-3.132
SUCACY	-0.01329	-0.006580	0.04246	0.02972
	-2.205	-1.024	0.1369	4.155
SUWHE	1.661	-0.1283	16.24	1.714
	2.889	-0.1514	0.5266	3.114
SUWHEP	-31.44	37.85	-277.7	-25.79
	-2.425	1.953	-0.3902	-1.978
SUWHEY	0.01436	0.01868	-0.4795	0.002812
	1.271	1.701	-0.7865	0.2839
ONE	-309.9	141.1	5417.	173.8
	-1.758	0.7106	0.5735	1.333
MPE	6252.	-12070.	-81460.	-11930.
	1.493	-2.594	-0.3539	-3.840
INCOME	9.289	8.582	13.79	0.8496
	3.336	2.666	0.08831	0.4075
NHSLDMEM	71.04	79.40	-867.2	84.22
	4.151	4.045	-0.8818	6.497
DAYS	4.841	10.44	91.66	10.07
	2.666	5.690	0.8265	5.251
H1	42.96	19.80	4200.	28.46
	1.630	0.6631	3.001	1.185
H2	70.77	65.24	-1044.	-8.614
	2.999	2.528	-0.8176	-0.5659
H3	5.457	-105.8	2165.	-6.041
	0.1808	-3.289	1.265	-0.2553
R^2	0.3430	0.5093	0.02489	0.4846
Obsevations	733	677	734	686
Standard error	669.2	734.4	36820.	468.0

Table 6.5c
Electricity demand model—Off-Season

Variable	March	April	May	October
SUSHE	0.1007	0.5632	0.7067	-0.04551
	0.7312	5.888	2.134	-0.8272
SUSHEP	0.003324	-0.03387	0.2622	0.1208
	0.2169	-4.198	3.186	3.238
SUSHEY	0.006488	0.005812	-0.04139	0.003601
	1.865	2.371	-3.599	2.471
SUCAC	-1.403	-0.2077	0.7335	-1.173
	-0.8885	-0.6318	3.177	-1.446
SUCACP	-0.1852	-0.1950	0.08567	0.05117
	-0.4703	-3.170	2.132	0.4563
SUCACY	0.02726	0.08173	-0.007448	0.05366
	0.5821	6.812	-0.8363	1.616
SUWHE	4.643	3.141	1.372	1.784
	3.947	6.821	2.112	3.867
SUWHEP	-57.39	-40.63	-7.829	-22.00
	-2.173	-4.237	-0.4863	-2.134
SUWHEY	0.01289	-0.01331	0.01608	0.002791
	0.7555	-1.429	1.355	0.3019
ONE	-281.5	-589.3	-384.3	66.22
	-0.8650	-4.696	-2.140	0.5137
MPE	-4175.	3236.	694.3	-9553.
	-0.5964	1.036	0.1652	-3.072
INCOME	8.132	3.721	8.199	5.094
	2.365	1.760	2.862	2.500
NHSLDMEM	37.76	49.32	66.09	96.10
	1.599	3.652	3.852	7.301
DAYS	11.59	11.04	6.273	6.648
	2.227	6.780	3.298	4.140
H1	283.0	-27.87	40.32	118.3
	4.375	-1.069	1.402	5.152
H2	32.10	0.5656	10.83	-5.837
	1.299	0.03618	0.5386	-0.3884
H3	-177.4	-19.63	-64.58	-13.37
	-3.621	-0.8231	-2.078	-0.5421
R^2	0.6350	0.7318	0.3613	0.5172
Observations	455	612	650	782
Standard error	686.0	472.4	624.0	520.4

Table 6.6a
Natural gas demand model—Winter

Variable	November	December	January	February
SUSHG	-0.00502	0.4425	0.4864	0.6028
	-0.06020	5.206	5.728	7.479
SUSHGP	0.01479	-0.02919	-0.05784	-0.08690
	5.901	-4.817	-6.112	-9.434
SUSHGY	0.00635	0.00637	0.00669	0.00988
	2.326	2.472	2.470	3.739
SUWHG	6.423	3.493	3.255	-10.10
	2.394	1.014	0.9789	-3.572
SUWHGP	-7.288	-3.866	5.780	42.42
	-1.198	-0.3346	0.5256	4.611
SUWHGY	0.03068	0.05402	-0.1333	0.08079
	0.4864	0.7748	-2.262	1.544
ONE	50.36	-42.89	-82.00	167.5
	1.144	-0.6622	-1.472	3.889
MPG	-210.2	-29.26	-28.10	-629.6
	-1.092	-0.1047	-0.1237	-3.592
INCOME	-1.311	-2.495	0.4447	-3.195
	-1.405	-1.936	0.4016	-3.673
NHSLDMEM	-7.059	3.817	2.567	0.7171
	-1.583	0.7559	0.6227	0.2098
DAYS	2.707	3.670	4.607	2.659
	3.465	5.126	7.012	3.724
H1	10.12	4.570	3.285	-5.157
	1.008	0.3746	0.3612	-0.6605
H3	-22.18	-22.17	-18.32	-13.10
	-2.250	-1.968	-1.944	-1.660
R^2	0.8709	0.7431	0.6939	0.6522
Observations	365	355	343	345
Standard error	105.3	121.2	90.07	79.25

The price interaction term SUSHEP is defined as the product of SUSHE with the marginal cost of comfort SHP. On the other hand, SHP is the product of the change in kwh, DSHE, and the marginal price of electricity MPE. Thus the price derivative of SUSHEP produces SUSHE × DSHE as indicated in the elasticity formula. The elasticity of total energy usage with respect to income is

Table 6.6b
Natural gas demand model—Summer

Variable	June	July	August	September
SUSHG	0.9063	0.7620	-0.07526	0.7513
	2.044	7.193	-0.4280	5.704
SUSHGP	-0.00133	-0.03000	0.04950	0.002045
	-0.01347	-2.565	4.066	0.2143
SUSHGY	-0.02381	0.008236	0.01043	-0.01296
	-1.881	3.328	2.006	-3.386
SUWHG	-3.715	0.9190	1.850	0.2676
	-2.828	1.002	1.417	0.1246
SUWHGP	15.32	0.9982	5.923	-1.276
	3.155	0.3537	1.386	-0.2042
SUWHGY	0.05563	-0.00402	-0.08782	0.01350
	2.826	-0.2351	-3.958	0.3393
ONE	59.25	7.050	-39.86	-5.085
	2.515	0.3926	-1.779	-0.1486
MPG	-259.0	-76.85	-190.2	-146.6
	-2.646	-1.141	-2.152	-1.107
INCOME	-0.1343	0.3778	1.862	0.9519
	-0.2709	0.9448	4.311	1.539
NHSLDMEM	0.3080	2.545	3.455	-0.05463
	0.1535	1.493	1.948	-0.02298
DAYS	0.3133	0.02302	1.100	1.232
	1.008	0.09119	3.505	2.787
H1	2.337	4.134	7.697	2.382
	0.5439	1.151	2.114	0.4672
H3	-2.369	-12.96	-10.09	0.6339
	-0.5549	-3.733	-2.656	0.1173
R^2	0.3206	0.8511	0.7433	0.5758
Observations	368	345	353	360
Standard error	47.70	39.04	41.17	53.34

$$(\text{INCOME}/\text{QUAN}) \, (\partial \text{QUAN}/\partial \text{INCOME}) \; = \; (\text{INCOME}/\text{QUAN}) \, \times$$

$$[\, (\text{SUSHE}) \, \gamma_2 + (\text{SUCAC}) \, \gamma_5 + (\text{SUWHE}) \, \gamma_7 + \gamma_{11} \,]$$

We evaluate the elasticities at the sample average values for the variables in a given billing period. We interpret these elasticities as short-run indicators of price and income responsiveness because they do not include the effect of price changes on portfolio compositions.

Table 6.6c
Natural gas demand model—Off-Season

Variable	March	April	May	October
SUSHG	0.5985	0.4169	0.2373	0.2712
	2.918	3.510	2.240	3.153
SUSHGP	-0.09167	-0.06732	0.07501	-0.001710
	-3.629	-3.292	4.521	-0.2561
SUSHGY	0.00095	0.00948	-0.00162	0.00301
	0.1403	3.170	-0.3595	1.085
SUWHG	-3.136	-1.561	-0.5396	2.197
	-0.6556	-0.8688	-0.6526	1.063
SUWHGP	12.03	12.79	-0.7165	-3.768
	0.7189	2.113	-0.3099	-0.6357
SUWHGY	0.1648	0.01130	0.03900	0.01825
	1.925	0.4741	2.747	0.4313
ONE	87.57	52.02	14.77	-50.36
	1.121	1.839	0.9288	-1.613
MPG	-275.6	-173.5	-98.63	-215.5
	-0.9145	-1.574	-1.524	-1.653
INCOME	-2.694	-0.9628	0.06898	0.00927
	-1.836	-1.693	0.2026	0.01339
NHSLDMEM	2.536	2.391	1.441	-0.04289
	0.6951	1.098	0.8760	-0.01504
DAYS	1.736	0.4740	0.8594	3.619
	2.084	1.548	3.690	7.967
H1	7.750	5.714	2.469	23.00
	0.6474	1.201	0.6502	3.461
H3	-11.18	-12.63	-4.711	-3.749
	-1.302	-2.628	-1.349	-0.5886
R^2	0.3728	0.4420	0.8432	0.7989
Observations	221	330	349	393
Standard error	58.76	50.94	37.92	72.37

To explain the sources of price and income sensitivity, we make a separate calculation of the "individual" price elasticity (MPE/QUAN) × γ_{10} as well as the elasticities for water heating, central air-conditioning, and space heating.

In the winter period, the relatively largest fraction of price and income sensitivity comes from water heating and other end-uses. Space-heating elasticities are relatively small while air-conditioning effects are not relevant. The total elasticities are well within the usual range of

Table 6.7a
Elasticities by billing period and HVAC system—Winter

Price Effect / Income Effect	November	December	January	February
Individual	-0.230 / 0.059	-0.138 / 0.055	-0.236 / 0.137	0.156 / 0.075
Water Heating	-0.077 / 0.026	-0.288 / 0.041	-0.212 / 0.016	-0.405 / 0.081
Central Air	-0.000 / 0.017	-0.000 / 0.002	0.000 / 0.004	0.000 / -0.013
Space Heating	0.003 / -0.070	0.010 / -0.004	-0.006 / -0.048	-0.017 / 0.010
Total	-0.304 / 0.031	-0.416 / 0.094	-0.454 / 0.109	-0.266 / 0.153

Table 6.7b
Elasticities by billing period and HVAC system—Summer

Price Effect / Income Effect	June	July	August	September
Individual	0.192 / 0.168	-0.353 / 0.144	-1.163 / 0.118	-0.442 / 0.018
Water Heating	-0.146 / 0.039	0.164 / 0.046	-0.632 / -0.656	-0.135 / 0.009
Central Air	0.043 / -0.037	0.021 / -0.017	0.023 / 0.059	-0.018 / 0.070
Space Heating	-0.001 / 0.025	-0.000 / -0.006	0.000 / 0.160	0.002 / 0.016
Total	0.089 / 0.195	-0.168 / 0.167	-1.771 / -0.318	-0.594 / 0.113

Table 6.7c
Elasticities by billing period and HVAC system—Off-Season

Price Effect / Income Effect	March	April	May	October
Individual	-0.139 / 0.168	0.117 / 0.079	0.026 / 0.176	-0.358 / 0.113
Water Heating	-0.257 / 0.036	-0.232 / -0.045	-0.043 / 0.050	-0.124 / 0.009
Central Air	-0.000 / 0.009	-0.004 / 0.083	0.011 / -0.015	0.000 / 0.016
Space Heating	0.001 / 0.053	-0.008 / 0.034	0.005 / -0.048	0.005 / 0.034
Total	-0.395 / 0.266	-0.128 / 0.152	-0.001 / 0.161	-0.476 / 0.172

estimated energy usage elasticities. In the summer period, price and income responsiveness are less precisely determined. Sensitivity arises from water heating and other end-uses. The air-conditioning price elasticities are estimated to be positive for three of the four months, but the combined effects are negative. Overall there appears to be lower price and greater income sensitivity in the summer compared with winter months.

Table 6.8a
Elasticities by billing period and HVAC system—Winter

Price Effect / Income Effect	November	December	January	February
Individual	-0.182 / -0.116	-0.019 / -0.163	-0.021 / 0.032	-0.582 / -0.303
Water Heating	-0.132 / 0.057	-0.054 / 0.076	0.081 / -0.184	0.719 / 0.141
Space Heating	0.041 / 0.167	-0.081 / 0.156	-0.151 / 0.171	-0.205 / 0.252
Total	-0.272 / 0.108	-0.154 / 0.069	-0.091 / 0.019	-0.069 / 0.089

Table 6.8b
Elasticities by billing period and HVAC system—Summer

Price Effect / Income Effect	June	July	August	September
Individual	-1.078 / -0.057	-0.295 / 0.150	-0.712 / 0.709	-0.434 / 0.289
Water Heating	1.354 / 0.499	0.082 / -0.034	0.444 / -0.670	-0.069 / 0.076
Space Heating	-0.000 / -0.148	-0.007 / 0.073	0.021 / 0.114	0.002 / -0.272
Total	0.276 / 0.295	-0.220 / 0.189	-0.247 / 0.153	-0.502 / 0.093

Table 6.8c
Elasticities by billing period and HVAC system—Off-Season

Price Effect / Income Effect	March	April	May	October
Individual	-0.348 / -0.391	-0.366 / -0.209	-0.319 / 0.023	-0.311 / 0.001
Water Heating	0.238 / 0.376	0.512 / 0.047	-0.049 / 0.268	-0.109 / 0.054
Space Heating	-0.127 / 0.020	-0.098 / 0.194	0.044 / -0.019	-0.003 / 0.069
Total	-0.236 / 0.005	0.048 / 0.032	-0.324 / 0.271	-0.423 / 0.126

The off-season period is characterized by income sensitivity in both air conditioning and space-heating. However, price sensitivity is again determined by water heat and other uses. The magnitudes of total price and income effects are similar to those obtained in the winter period. Electricity demand is estimated to reveal greatest price sensitivity in September and largest income elasticity in June.

Estimation of natural gas demand indicates larger price effects and somewhat smaller or equal-size income effects as compared with electricity demand. A striking feature of the elasticities in Table 6.8 is that much greater price sensitivity may be attributed to space heating. In the summer period, price sensitivity arises from the combination of water heating and other uses. Again, the off-season appears more like the winter in terms of price and income elasticities. The largest price sensitivity for natural gas demand occurs in the off-season while the largest income sensitivity occurs in the summer.

The elasticities calculated in Tables 6.7 and 6.8 do not account for substitution of HVAC systems in the long run. In the next section we illustrate the long-run effects by analyzing once again the consequences of the ASHRAE mandatory thermal standards.

6.5. Effects of ASHRAE standards on energy demand

In this section we calculate electricity and natural gas demand under alternative mandatory energy conservation policies proposed by ASHRAE. The effects of these policies on HVAC-system choice was considered in Chapter 3. Recall that the six cases to consider are in turn: (1) baseline scenario, (2) increased wall and ceiling insulation, (3) relaxed design temperatures, (4) stormed windows and sealed air cracks, (5) modified thermostat settings, and (6) employment of a combination of policies two through four to achieve maximal conservation.

For long-term forecasting we do not require predictions at the monthly level. Instead, we estimate equation (5) using annual data formed from the monthly billing data. The thermal model is used to provide annual estimates of heating and cooling unit energy consumptions. Variable means for the annual model are given in Table 6.9 while regression results are given in Table 6.10. The results of the regression analysis are in accord with the theory and with the results obtained using the billing data. Coefficients are generally significant in the electricity demand model and indicate positive income and negative price effects. In the natural gas model, the estimated coefficients are again quite significant but the water-heating variables appear with anomalous price and income effects. Interestingly, the selectivity correction terms are not significant in the annual models, which suggests that the aggregation masks the correlation effect found in the billing data.

Table 6.11 presents elasticities calculated at sample means based on the annual data. The elasticity estimates are comparable to those obtained using the billing data. Overall, water heating and other uses provide the greatest price sensitivity in the electricity model while both price and income sensitivity arise principally from space heating usage in the natural gas demand model.

Table 6.9
Variable means annual model

SUSHE	5311.	SUSHG	1436.
SUSHEP	258100.	SUSHGP	30570.
SUSHEY	135800.	SUSHGY	36010.
SUCAC	325.	-	-
SUCACP	1523.	-	-
SUCACY	8676.	-	-
SUWHE	1528.	SUWHG	158.6
SUWHEP	54.45	SUWHGP	36.34
SUWHEY	36480.	SUWHGY	3857.
MPE	0.04005	MPG	0.2380
INCOME	23.0	INCOME	23.2
NHSLDMEM	3.26	NHSLDMEM	3.24
QUAN (kwh)	13000.	QUAN (therms)	1186.
QEBASE (kwh)	4847.	QGBASE (therms)	105.2
DSHE	267.7	DSHG	68.0
SHP	8.80	SHP	15.33
SHE	0.2086	SHG	0.5895
DCAC	55.7	-	-
ACP	1.20	-	-
CAC	0.3513	-	-
Observations	911		655

With the specification and estimation of the annual model completed, we proceed to the simulations. Recall that the period for the simulation begins in 1978 and ends in the year 2000. Our price forecast assumes that the price of natural gas doubles during this period while the price of electricity grow somewhat less rapidly. The policy simulations in Chapter 3 indicate that electric forced-air and electric baseboard will show increased penetration while gas space-heating systems will decline. Conservation policies were seen to emphasize these trends while causing further shifts from electric forced-air to baseboard systems. The conservation policies affect demand in two ways. First, expected demand depends on the probability of HVAC-system choice. Second, the conservation policies affect the predicted unit energy consumption levels in the thermal model.

Taking expectations in equation (4) yields

$$E(Y_{it}^f) = Z_i\gamma + UEC_{it}^{SH} P_i^{SH} (X_{it}^{SH}\beta^{SH}) + UEC_{it}^{AC} P_i^{AC} (X_{it}^{AC}\beta^{AC})$$
$$+ UEC_{it}^{WH} P_i^{WH} (X_i^{WH}\beta^{WH}) + X_{it}^M\beta^M \qquad (6)$$

We evaluate equation (6) using the estimated parameters from the annual regression models. Each conservation policy affects the unit

Table 6.10
Annual electricity and natural gas regressions

SUSHE	0.3909 6.816	SUSHG	0.5694 7.592
SUSHEP	-0.002159 -3.638	SUSHGP	-0.009291 -7.282
SUSHEY	0.002508 1.899	SUSHGY	0.009522 4.767
SUCAC	1.921 2.127	-	- -
SUCACP	-0.1891 -1.842	-	- -
SUCACY	.07398 3.153	-	- -
SUWHE	1.582 3.139	SUWHG	0.5173 0.4586
SUWHEP	-16.80 -1.559	SUWHGP	7.363 2.157
SUWHEY	0.01227 1.506	SUWHGY	-0.04546 -2.014
ONE	869.2 0.7929	ONE	267.6 1.441
MPE	-13830. -0.5897	MPG	-985.4 -1.627
INCOME	31.94 1.662	INCOME	-2.818 -0.9311
NHSLDMEM	712.0 6.186	NHSLDMEM	18.92 1.284
H1	580.4 1.772	H1	27.23 0.8831
H2	20.36 0.1206	-	- -
H3	-311.1 -1.198	H3	-58.75 -1.533
R^2	0.6963		0.6038
Observations	911		655
Standard error	4913.		476.9

energy consumption terms as well as the predicted probabilities. We calculate the unconditional probabilities for water-heating fuel, space-heating fuel, and central air conditioning using the nested logit specifications estimated in Chapter 3. The assumed growth in energy prices influences both the probabilities through operating costs and the forecasted fuel usage through the utilization terms. Results of the forecasts are presented in Tables 6.12 and 6.13 for electricity and natural gas respectively.

Table 6.11

Elasticities in annual model by HVAC system

Price Effect / Income Effect	Electricity	Natural Gas
Individual	-0.043 / 0.057	-0.198 / -0.055
Water Heating	-0.079 / 0.033	0.234 / -0.141
Central Air	-0.011 / 0.002	
Space Heating	-0.009 / 0.024	-0.182 / 0.267
Total	-0.142 / 0.116	-0.146 / 0.072

Table 6.12

The effect of ASHRAE thermal policies on electricity demand

Policy	1978	1985	1990	2000
1. Baseline	1.000	1.005	1.011	1.025
2. Wall and ceiling insulation	0.964	0.969	0.974	0.989
3. Design temperatures	1.003	1.009	1.014	1.029
4. Window treatment and infiltration	1.004	1.009	1.016	1.032
5. Thermostat setting	0.950	0.954	0.959	0.973
6. Policies 2-4	0.970	0.977	0.983	0.999

Table 6.13

The effect of ASHRAE thermal policies on natural gas demand

Policy	1978	1985	1990	2000
1. Baseline	1.000	0.911	0.833	0.679
2. Wall and ceiling insulation	0.855	0.781	0.715	0.581
3. Design temperatures	1.011	0.918	0.838	0.679
4. Window treatment and infiltration	1.004	0.917	0.841	0.688
5. Thermostat setting	0.875	0.802	0.738	0.607
6. Policies 2-4	0.866	0.789	0.721	0.585

Average demand for the baseline scenario in 1978 is normalized to unity so that all estimates are given relative to this period. In the baseline scenario, electricity demand is seen to increase. This is not surprising, as we have assumed that the relative price of electricity to natural gas is declining in real terms. The increased usage in electricity arises solely from portfolio shifts that increase the share of electric space- and water-heating systems.

Increased wall and ceiling insulation lowers the utilization of electricity but does not reverse the trend toward greater electrification. Moderated design temperatures and effective window and infiltration treatment raise usage relative to the baseline scenario. Decreased winter and increased summer thermostat settings produce the greatest response in terms of lowering electricity demand. The maximal conservation scenario, which combines wall and ceiling insulation improvements with moderated design temperatures and effective window treatments, is seen to result in quick reductions in electricity that grow to the 1978 baseline values by the year 2000.

The demand for natural gas falls inversely with the projected forecasts for electricity usage. Under the baseline scenario, natural gas usage falls to 68 percent of its 1978 baseline level. A policy of moderate thermostat levels leads to lower penetration of gas heating systems and lower usage levels than under the increased insulation scenario. Under maximal conservation and rising relative gas prices, usage falls to 59 percent of baseline values.

In summary, we have examined a model of residential electricity and natural gas demand using a pooled cross-sectional time-series of microdata on individual households. The objective has been to provide more precise explanations of household behavior than have been previously available from studies using annual consumption data. We have been able to predict seasonal variation in fuel usage demand using a combined engineering economic model.

The availability of highly disaggregated data provides an opportunity to explain more detailed, subtle variation in consumer behavior than is possible with aggregate data. Indeed, we have seen that the endogeneity of the appliance choice decision is a seasonal phenomenon suggesting that households at some times behave as if the appliance stock is exogenous, but at other times behave according to the underlying selection rules that match usage and system types.

We have also been able to determine the composition of short-run elasticities and have found that fuel price sensitivity varies both seasonally and by equipment types. Finally, we have investigated the long-run elasticity of fuel utilization by including the effect of portfolio shift as well as price and income substitution. The next step in appliance choice and energy demand analysis should be to incorporate these results into existing microforecasting models.

A review of the National Interim Energy Consumption Survey billing data

A.1. Introduction

This appendix reviews the National Interim Energy Consumption Survey (NIECS) billing data and presents the procedures used to process the data into a form useful for econometric research. The NIECS data contain detailed energy demand information at the household level for 4081 households over the period April 1978 to March 1979. Data include the structural characteristics of the housing unit, demographic characteristics of the household, fuel usage, appliance characteristics, and actual energy consumption over the 12-month period. The NIECS annual file coded 59 separate variables to report these items. The annual data is summarized in Table A.1.

The preparation of the NIECS annual file was undertaken by Thomas Cowing, Jeffrey Dubin, and Daniel McFadden in the summer of 1981. At that time an evaluation of the data was made to determine its usefulness for an energy demand study. For details concerning this evaluation, the reader may consult Cowing, Dubin, and McFadden (1982). In their report, Cowing, Dubin, and McFadden review the NIECS data and assess the problems of measurement error, sample design, and missing data imputation.[1]

A collateral evaluation of the NIECS data was conducted by Carl Blumstein, Carl York, and William Kemp (1979). This report has been reviewed and evaluated by Cowing, Dubin, and McFadden (1982). Independent reports on the weather information contained in the NIECS data set and on procedures used to find state locations for NIECS households are given in Cowing, Dubin, and McFadden (1982). Finally, a collection of source programs and documentation related to the processing and creation of the NIECS/PNW data base is available in Dubin (1983).

---

[1]Related source documents are Response Analysis Corporation (1981) and U.S. Department of Energy (DOE/EIA- 0272, 0272/5, 0272/2, 0272/3, 1017/5, 0262/1, 0262/2, 0207/1).

Table A.1

A summary of NIECS information[a]

Housing characteristics	*Heating/cooling equipment*
Housing type	Main heating system type and fuel
Year house built	Secondary heating system type and fuel
Number of floors	Type of air conditioning equipment
Floor area	Number of rooms air-conditioned
Number of rooms	
Number and type of windows	*Household appliances*
Number and type of stormed windows	Fuel used for water heating
Number and type of outside doors	Number and type of refrigerators
Number of stormed doors	Number and type of cooking equipment
Presence, type, amount of attic	Use of other household appliances
insulation	
Wall insulation	*Demographic characteristics*
	Number age, sex, and employment
Retrofit/conservation efforts[b]	status of household members
Stormed windows	Marital status of respondent
Weatherstripping	Race of respondent
Clock thermostat	Education of respondent and spouse
Attic insulation	Total household income for 1977
Wall insulation	Housing tenure (own or rent)
Floor insulation	
Water pipe insulation	*Energy use and consumption*[c]
Water heater insulation	Use of electricity, natural gas,
Other insulation	LPG, and fuel oil
Caulking	- for different functions
Plastic coverings on windows	- paid by household
or doors	- consumption and expenditure
Other information	
Geographic location	
Heating degree-days	
Cooling degree-days	
Type of community	

[a]Questions were also asked about ownership and use of motor vehicles, but this information was not relevant to this project.

[b]Refers to conservation actions taken between January 1977 and the date of the interview, fall 1978.

[c]Data on monthly household fuel consumption and expenditures by type of fuel were obtained from fuel suppliers. The data cover the one-year period from April 1978 through March 1979.

Section A.2 considers the NIECS billing data and reviews procedures used to reprocess the data in a form suitable for econometric research. Section A.3 examines the use of the monthly billing data in the construction of seasonal marginal prices. Section A.4 considers a case study of a particular NIECS household as an illustration of the data structure and as a detailed internal consistency check.

A.2. Processing the monthly billing data

In this section we discuss the monthly billing data matched to the National Interim Energy Consumption Survey (NIECS). Following a brief review of the data collection procedure, we describe our strategy to process the raw billing data into a form useful for econometric analysis. Summary information based on the processed data provides a measure of data quality for empirical studies.

NIECS is a four-stage area probability sample consisting of 103 primary sampling units. The NIECS sample was drawn from the contiguous United States and the District of Columbia. In final form the sample represents individually specific information on 4081 households. In 3842 cases, demographic and structural attributes were obtained by personal interview. In the remaining 239 cases, data were obtained by mailed questionnaire, and the contractor, Response Analysis Corporation, found it necessary to impute a substantive number of missing responses. At the completion of each interview, households were asked to sign a Department of Energy waiver allowing Response Analysis Corporation to collect data on fuel utilization directly from the appropriate fuel supplier. Utilities responded in varying degrees of completeness. Table A.2 summarizes the data collection response rates for 4080 households that used electricity. Referring to Table A.2, we see that in approximately three-fourths of the sample, at least 11 months of billing data were collected. This is a strikingly high percentage of the cases. In an additional 12 percent of the sample, fewer than ten months of billing data were collected. For these households, the contractor provided imputed annual information using various "hot-deck" and regression estimates. The usefulness of the imputed annual figures for econometric analysis seems questionable, so it would seem best to concentrate empirical efforts on the first group with nearly complete data.

For each household, a maximum of 20 billing periods were recorded with an average length of 30 days per billing period. In each billing period the following information was recorded: the expenditure in dollars for the fuel, the quantity in kilowatt-hours for electricity consumed, the beginning year, month, and date, and the ending year, month, and date. Also recorded were a code for whether or not the beginning and ending dates were known or imputed, whether the end of each billing period was an actual or estimated meter reading, and the total number of heating and cooling degree-days for the billing period computed to 14 separate bases.

Table A.2

Energy consumption records and missing data
for survey households using electricity

	Number of households	Percent
Total households using fuel	4080	100.0
Data received from fuel supplier	3509	86.0
11 months or more	3023	74.1
5-10 months	340	8.3
Less than 5 months	146	3.6
Household pays directly to supplier—no data available	334	8.2
Household not identified in company records	128	3.1
Company refused to participate	0	0
Company unknown or not located	0	0
Authorization form not signed	206	5.1
Fuel used included in rent or paid in other way	237	5.8

Source: *NIECS: Report on Methodology*, Part I, Section 5. Household and Utility Company Surveys. Princeton, N.J.: Response Analysis Corporation, February 1981.

In all cases the month in which the billing period took place was known with certainty. Documentation provided by Response Analysis Corporation indicates that there were two major categories of billing date completeness.

The first category consists of the majority of dates unknown for all billing periods. In this case, billing periods were assumed to begin on the fifteenth of the month and end on the fifteenth of the following month, with the beginning and ending date codes set to indicate that this assumption had been made. The second category consists of households in which specific dates were unknown for only a few periods at the beginning of the billing record. In this case the initial months were assigned a billing date equal to the first known billing date. It is possible to determine the exact duration of a billing period for only those cases in which the beginning and ending dates are known with certainty.

Table A.3 exhibits the actual data from the NIECS billing tape for the ninetieth household. From Table A.3 we see that 14 billing periods were coded. Columns C and D indicate whether the beginning and ending dates are known (0) or unknown (1). As columns C and D consist of all 0's, we know that all dates for observation 90 were known with certainty. Reading across the top row of Table 3, we see that the starting date was

Table A.3
Raw billing data for observation #90

													0	8
403	38.28	0	0	78	1	19	78	2	23	35	1455	0	1	1
290	28.76	0	0	78	2	23	78	3	23	28	920	0	2	1
280	28.50	0	0	78	3	23	78	4	20	28	553	0	3	1
341	35.79	0	0	78	4	20	78	5	23	33	398	8	4	2
261	28.95	0	0	78	5	23	78	6	21	29	37	70	5	2
290	31.42	0	0	78	6	21	78	7	21	30	7	196	6	2
232	25.57	0	0	78	7	21	78	8	20	30	0	303	7	1
280	30.25	0	0	78	8	20	78	9	18	29	30	135	8	2
251	27.38	0	0	78	9	18	78	10	17	29	237	5	9	1
340	33.12	0	0	78	10	17	78	11	20	34	485	0	10	1
331	32.54	0	0	78	11	20	78	12	20	30	833	0	11	1
425	39.54	0	0	78	12	20	79	1	18	29	1020	0	12	1
303	29.29	0	0	79	1	18	79	2	24	37	1528	0	13	1
206	20.32	0	0	79	2	24	79	3	22	26	670	0	14	1
0	0	0	0	0	0	0	0	0	0	0	0	0	15	0
0	0	0	0	0	0	0	0	0	0	0	0	0	16	0
0	0	0	0	0	0	0	0	0	0	0	0	0	17	0
0	0	0	0	0	0	0	0	0	0	0	0	0	18	0
0	0	0	0	0	0	0	0	0	0	0	0	0	19	0
0	0	0	0	0	0	0	0	0	0	0	0	0	20	0
A	B	C	D	E	F	G	H	I	J	K	L	M	N	O

A:	Quantity kwh	I:	Ending month
B:	Expenditures in Dollars	J:	Ending day
C:	Beginning date known	K:	Elapsed days
D:	Ending date known	L:	Heating degree-days—65°F
E:	Beginning year	M:	Cooling degree-days—65°F
F:	Beginning month	N:	Billing period number
G:	Beginning day	O:	Ending of period:
H:	Ending year		actual or estimate code

January 19, l978 (columns E, F, and G), and that the ending date was February 23, 1978 (columns H, I, and J), which corresponds to 35 elapsed days (column K). Quantity, expenditure, and heating and cooling degree-days (base 65) are recorded in columns A, B, L, and M respectively.

In the econometric analysis of the demand for electricity, we must ensure that all observations correspond to the behavior of economic agents. Thus we follow a procedure for processing the raw data that determines quantities and expenditures for periods of time bounded at each end by actual meter readings. The estimated versus actual code is given in column O of Table A.3. The codes in this case are 0 for no data, 1 for actual meter reading, 2 for estimated reading, 8 for no information provided from utility on this item, and 9 for fuel not used. Note that these codes refer to the end of the period, so that it is impossible to tell whether period one data is actual or estimated.

Given the possibility that a code 8 corresponds to an actual meter reading rather than an estimated reading, we have followed the convention of bounding observations by code 1's in place of code 8's and flagging the later cases to indicate their suspect quality. Given that we do not have any information from the utility for the beginning of period one (i.e., the end of period zero), it would seem useful to treat the end of period zero as if it had been assigned with a code 8.

Reading down column O, we see that the end of period zero, i.e., the beginning of period one, has the assigned 8 code. In the next line we see that the end of period one corresponds to an actual meter reading. Thus billing period one provides a *tentatively* valid observation. Comparing the rows of Table A.3 for billing periods one and two, we see that the end of period one (equivalent to the beginning of period two) is an actual meter reading. Also, the end of period two is an actual reading, so that billing period two is bounded by actual readings.

As we go down the table, we see that the beginning of period four is an actual reading, but that the ends of periods four, five, and six are estimated. Not until the end of period seven do we have another actual reading. We thus aggregate the information in periods four, five, six, and seven to obtain a single observation bounded at each end with actual meter readings. This aggregated period contains 1124 kilowatt-hour consumption (341 + 261 + 290 + 232) and corresponds to 122 days, or approximately four months.

We wrote a FORTRAN program that processes the raw billing data and produces the following variables: flag code (defined in Table A.4), start code, end code, expenditure, heating and cooling degree-days (base 65 and base 75), and quantity consumed.

A 0 value for the flag code indicates no data, a 1 indicates that the processed observation is bounded by actual meter readings, a 2 indicates actual meter readings at both ends of the period but at least one imputed date at either end point, a 3 indicates that at least one end point corresponds to the 8 code (no information on type of meter reading), and a 4 idicates that it is not possible to determine whether one of the end points is actual or estimated and that at least one end point has an imputed date.

Table A.5 illustrates the processed data for observation 90. In our processing we found it adequate to allow space for up to 15 billing

Table A.4

Explanation for variable flag

Code	Definition
0	No data
1	Actual meter readings; known dates
2	Actual meter readings; at least one imputed date
3	No data on actual versus estimated; known dates
4	No data on actual versus estimated; at least one imputed date

Table A.5

Processed billing data for observation #90

Flag	Start code	End code	Expenditure	HDD	CDD	Quantity
3.00	18.00	53.00	38.28	1455.00	0.0	403.00
1.00	53.00	81.00	28.76	920.00	0.0	290.00
1.00	81.00	109.00	28.50	553.00	0.0	280.00
1.00	109.00	231.00	121.73	442.00	577.0	1124.00
1.00	231.00	289.00	57.63	267.00	140.0	531.00
1.00	289.00	323.00	33.12	485.00	0.0	340.00
1.00	323.00	353.00	32.54	833.00	0.0	331.00
1.00	353.00	382.00	39.54	1020.00	0.0	425.00
1.00	382.00	419.00	29.29	1528.00	0.0	303.00
1.00	419.00	445.00	20.32	670.00	0.0	206.00
0.00	0.00	0.00	0.00	0.00	0.0	0.00
0.00	0.00	0.00	0.00	0.00	0.0	0.00
0.00	0.00	0.00	0.00	0.00	0.0	0.00
0.00	0.00	0.00	0.00	0.00	0.0	0.00
0.00	0.00	0.00	0.00	0.00	0.0	0.00

periods rather that the 20 records allowed for in the raw data. Note that while Table A.3 reports information on 14 billing periods, the processed information corresponds to ten observations in Table A.5. The start and end codes summarize the seven variables allocated in the raw data set for beginning and ending dates and elapsed days. These codes are defined as the number of days since January 1, 1978. A negative number would thus correspond to a number of days before January 1, 1978. The difference between the start and end codes for any billing period is then the elapsed number of days. For example, the start code in Table A.5 for the first processed observation indicates that it begins 18 days past the first of January for an elapsed time of 35 days. This number may be cross-checked in Table A.3.

Note that the first three processed observations in Table A.5 are identical to their counterparts in Table A.3. The fourth observation in Table

A.5 corresponds to the aggregation of periods four, five, six, and seven from Table A.3. Finally, the flag code in the first column of Table A.4 is appropriately set for each processed observation as can be checked with the aid of Tables A.3 and A.4.

Table A.6 provides a summary of the processing of 2018 cases for which the certainty code of housing location match was greater than three and for which the household was an owner-occupied, single-family, detached dwelling.[2]

Table A.6

Summary statistics for variable flag: electricity billing data

Code	Absolute frequency	Relative frequency (pct)	Adjusted relative frequency
0	5,809	20.48	0
1	18,015	63.51	79.87
2	1,496	5.27	6.63
3	2,635	9.29	11.68
4	410	1.45	1.82

Total cases: 28,365, 127 missing, 1,891 partial

Table A.7 provides a similar summary for the processing of the natural gas billing data.

Table A.7

Summary statistics for variable flag: natural gas billing data

Code	Absolute frequency	Relative frequency (pct)	Adjusted relative frequency
0	4,827	28.13	0
1	10,869	63.34	88.13
2	122	0.71	0.99
3	1,195	6.96	9.69
4	147	0.86	1.19

Total cases: 17,160, 874 missing, 1,144 partial

Tables A.6 and A.7 indicate that information was unavailable for 127 households in the electricity data and unavailable for 874 households in

[2]For details on the location match, the reader may consult Cowing, Dubin, and McFadden (1982).

the natural gas data. However, 79.87 percent and 88.13 percent of the electricity and natural gas billing data are assigned a flag code of 1, which indicates a very high quality for the overall processed data sets.

A.3. Use of billing data to obtain marginal prices

This section considers the construction of seasonal marginal prices for electricity and natural gas from the monthly billing data. Details concerning the theory of this calculation (as opposed to its implementation) are presented in Chapter 4.

In the process described above of developing processed data from the raw monthly data, we emphasized the need to bound each observation by actual meter readings. These observations correspond to the behavior of the individual. In determining bills, however, it is likely that estimated as well as actual quantities are applied to the rate schedule by the utility. Thus to determine marginal price, we recommend the use of the billing data as it appears in the monthly data set.

Under the assumption that the rate schedule can be approximated by a two-part tariff, an appropriate procedure collects all observations within a primary sampling unit (this roughly corresponds to the area covered by a single utility) and fits a marginal price using ordinary least squares regression of expenditure on a constant term and quantity:

$$E_t = \alpha + \beta Q_t + V_t \tag{1}$$

with E_t = expenditure by observation t, Q_t = quantity consumed by observation t, V_t = random error term for observation t, α = fixed charge in two-part tariff, and β = marginal price. We now comment on issues of confidentiality, which introduce error in the billing data, and then remark on modifications to equation (1) that permit the estimation of seasonal prices.

A.3.1. Issues of confidentiality

Before public release of the NIECS billing data, a procedure designed to protect consumer confidentiality randomly adjusted the beginning and ending date of each billing period by up to three days. This procedure was designed to prevent matching of individual households with the billing data provided by the fuel supplier. Does this inoculation prevent recovery of marginal rates? Suppose we assume that the two-part tariff is an adequate representation of the billing schedule and that a random fraction ξ_{2t} of NIECS billing period two is assigned to NIECS billing period one to produce an (expenditure, quantity) observation (E_t^*, Q_t^*). Let (E_{1t}, Q_{1t}) and (E_{2t}, Q_{2t}) be the actual expenditure, quantity pairs for two contiguous billing periods determined by relation (1). Then:

$$E_t^* = E_{1t} + \xi_{2t} E_{2t} \qquad \text{and} \tag{2}$$

$$Q_t^* = Q_{1t} + \xi_{2t} Q_{2t} \tag{3}$$

From equation (1) :

$$E_{1t} = \alpha + \beta Q_{1t} + V_{1t} \qquad \text{and} \tag{4}$$

$$E_{2t} = \alpha + \beta Q_{2t} + V_{2t} \tag{5}$$

Thus:

$$E_t^* = \alpha + \beta Q_t^* + V_{1t} + \xi_{2t} V_{2t} + \alpha \xi_{2t} \qquad \text{so that} \tag{6}$$

$$E_t^* = \alpha + \beta Q_t^* + \varepsilon_t \qquad \text{where} \tag{7}$$

$$\varepsilon_t = V_{1t} + \xi_{2t} V_{2t} + \alpha \xi_{2t} \tag{8}$$

If ordinary least squares is an appropriate technique for estimation of (1), it should also provide consistent estimates of the parameters in (7). Thus, the inoculation done by Response Analysis Corporation would not appear to invalidate the basic statistical integrity of the procedure used to determine marginal prices, although it is expected that the standard error of the least squares regression will increase due to the noise introduced by the randomization process.

A.3.2. Issues of seasonality

We now consider a regression strategy that groups households within a primary sampling unit by season or other attributes that generally affect rate determination. We allow for the following possible rate schedules:

1 - all-electric home in the winter
2 - all-electric home in the summer
3 - all-electric home during the off-season
4 - non-all-electric home in the winter
5 - non-all-electric home in the summer
6 - non-all-electric home during the off-season

All-electric homes are households that have and use an electric space heating system. Winter is defined as billing periods that begin or end in January 1978. Summer is defined as billing periods which begin or end in July 1978. The off-season is defined as any billing period that begins in April, October, or September or ends in April, October, or November. The resultant partition closely matches the pattern exhibited by a significant majority of utilities during 1978.

In Figures A.1 through A.4, we plot expenditure versus quantity for selected NIECS locations. We use the symbols 1, 2, 3, 4, 5, and 6 to indicate the observation of a quantity-expenditure pair in a particular cell. It is possible for some cells to be empty (notably all-electric homes in some primary sampling units) so that not all points will be found in each figure. Finally, in each cell we have fitted a two-part tariff using least squares. Formal grouping tests are not presented as the figures are intended to illustrate the qualitative variety of rate schedules in the NIECS data and to suggest appropriate regression strategies for the estimation of marginal price.

In Figures A.1.1 and A.1.2, we see little evidence of seasonal structure. Figure A.1.1 indicates the possibility, however, that a winter rate may be distinguished from the rest of the seasons. (If one checks the national electric rate book for Newark, New Jersey, 1978, this supposition is verified.) In Figure A.2.1 we note that estimates of marginal price for all-electric households do not differ significantly from those of the non-electric homes. Furthermore, seasonality in rates is not exhibited on the basis of the slope estimates.

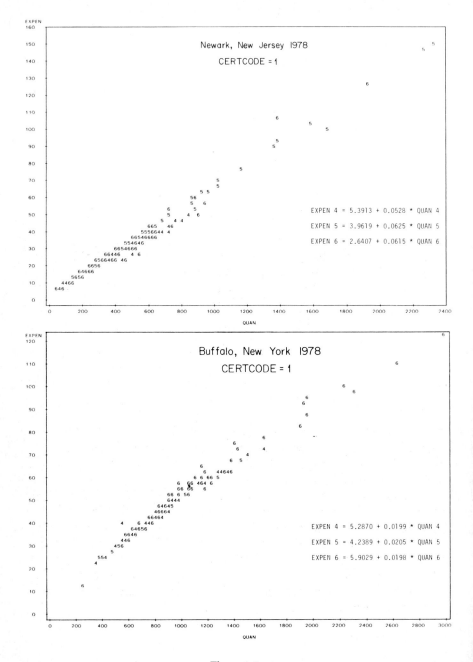

Figure A.1

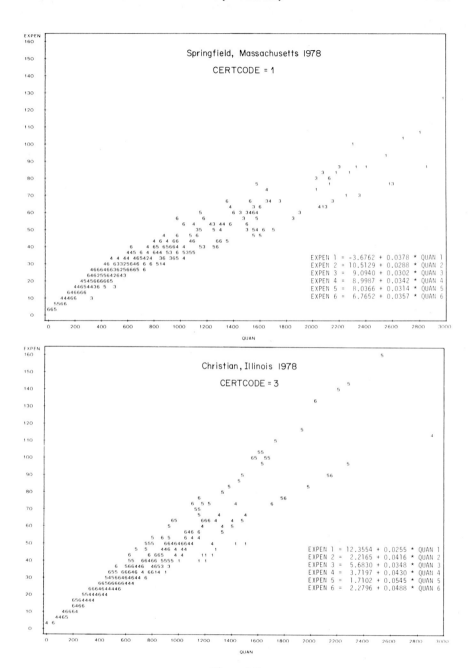

Figure A.2

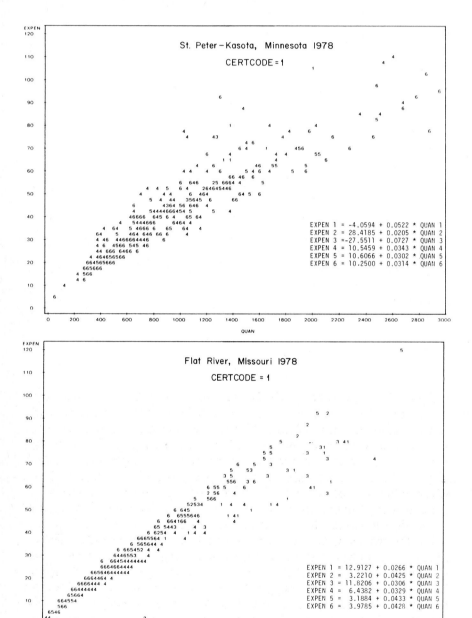

Figure A.3

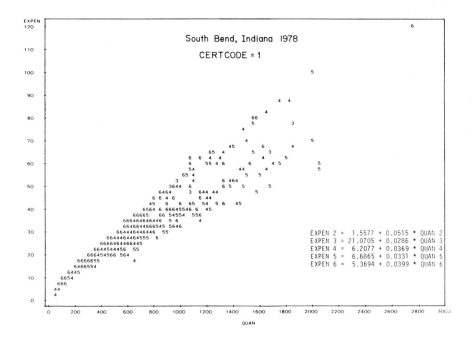

Figure A.4

Figure A.2.2 provides a striking illustration of multiple rate schedules. As we pass the 1400 kwh range, households in cells 5 and 6 (non-all-electric; summer and off-season) appear to fall on two distinct lines. Also, the slope estimates indicate a lower marginal price for all-electric homes as is illustrated by the households in cell 1, which tend to cluster below all other households. (Again we have consulted the rate books, which indicate multiple rates for small and large users of electricity in the Christian, Illinois, cluster.) Figure A.3.1 yields an imprecise picture for all-electric homes due perhaps to their small numbers. The price estimates for groups 4, 5, and 6 do not appear to be significantly different. Figure A.3.2 indicates some clustering of all-electric homes in cell 1 and the possibility of an all-electric rate. The winter rate for non-all-electric homes is lower than the estimated rates in cells 5 and 6. This does not provide an indication of a winter peaking rate. Finally, Figure A.4 shows a definite split in cluster 5 households, while the number of all-electric homes is too small to form any general conclusions.

In summary, we see that a two-part approximation to the rate schedule provides a qualitative tool to determine the presence of seasonal and differential rate schedules. Furthermore, when large numbers of observations are present the loss of efficiency from grouping observations into plausible rate cells is compensated for by avoiding basic specification bias.

The FORTRAN program that processes the raw electricity billing data constructs four marginal prices: AEMPE78—marginal price of electricity for all-electric homes, SMPE78—summer marginal price of electricity, WMPE78—winter marginal price of electricity, and OSMPE78—off-season marginal price of electricity in 1978. Consistency conditions and internal checks are imposed on the estimated prices so that at least ten observations are used in the regression analysis and so that winter and summer rates are in fact peak rates. A corresponding program to determine natural gas price by primary sampling unit does not attempt to discern a seasonal effect. For details the reader is referred to Dubin (1983).

A.4. Case study of household #1271

This section illustrates the processing of data on a selected household in the NIECS data file. The household was selected on the basis of three criteria: it is in Boston, Massachusetts (a location for which additional weather-related information was readily available), it can provide both electricity and natural gas billing data, and it uses or employs one of 19 alternative HVAC systems of particular interest to our study. The household selected is identified by a unique Department of Energy identification number: 1271.

Tables A.8 and A.9 present the processed billing data for household 1271. The electricity billing data cover a period of 462 days; the gas billing data cover a period of 394 days. Tables A.10 and A.11 present the thermal model output for electricity and natural gas respectively. The variable SHUEC refers to predicted heating usage in thousands of Btu's; ACUEC refers to predicted cooling usage in thousands of Btu's. The variables DSHUEC and DACUEC give the marginal increase in energy utilization for a 1-degree change in the thermostat setting for the period in question. In the heating mode this corresponds to raising ambient temperature from 70 to 71 degrees; in the cooling mode this corresponds to a change in temperature from 75 to 74 degrees.

Table A.8
Electricity billing data—household 1271

HHIDNO	FLAG	Start code	End code	QUAN	EXPEN	HDD65	CDD65	HDD75	CDD75
1271	3	-16	19	1199	61.95	1273	0	1633	0
1271	1	19	80	2026	97.24	2442	0	3072	0
1271	1	80	109	913	45.79	695	0	995	0
1271	1	109	139	719	37.98	470	0	780	0
1271	1	139	170	721	37.59	99	14	405	0
1271	1	170	201	832	41.23	22	119	225	2
1271	1	201	232	930	44.78	2	216	122	16
1271	1	232	261	760	37.83	100	35	365	0
1271	1	261	292	820	41.22	390	0	710	0
1271	1	292	321	830	40.91	510	0	810	0
1271	1	321	353	968	47.80	974	0	1304	0
1271	1	353	382	876	40.79	1062	0	1362	0
1271	1	382	410	815	35.08	1201	0	1491	0
1271	1	410	446	1059	44.65	1176	0	1546	0
0	0	0	0	0	0	0	0	0	0

Table A.12 presents the actual values of selected variables for household 1271. To compare the processed billing data with the annual information (including the thermal model output based on the annual data), we have selected a subset of the observations that corresponds to a period of approximately one year. These subsets lie within the dotted lines in Tables A.8, A.9, A.10, and A.11. Table A.13 presents the results of aggregating the billing data for the year. Note that ACUEC and DACUEC in Table A.13 have not been adjusted to reflect system coefficient of performance (COP), while similar variables in Table A.12 do reflect the COP adjustment. As may be seen by inspection, the estimates in Table A.13 compare favorably with each other and with those of the annual file (in Table A.12). Furthermore, the thermal model aggregates well across time and gives values that track the temperature profile closely.

The predicted thermal model coefficients (discussed in Chapter 2) imply that delivered energy per hour (Btuh) on a winter day with mean ambient temperature t and thermostat setting τ is:

$$Q = 2909.3 + 556.03\,(\tau-t) + 2.1305\,(\tau-t)^2$$

Table A.9
Natural gas billing data—household 1271

HHIDNO	FLAG	Start code	End code	QUAN	EXPEN	HDD65	CDD65	HDD75	CDD75
1271	3	87	119	171.20	68.79	706	0	1036	0
1271	1	119	178	128.80	59.73	391	37	964	0
1271	1	178	239	103.90	45.87	34	318	362	16
1271	1	239	300	175.20	75.92	593	23	1200	0
1271	1	300	361	271.80	111.62	1654	0	2284	0
1271	1	361	481	663.20	266.57	3914	0	5154	0
0	0	0	0	0	0	0	0	0	0
0	0	0	0	0	0	0	0	0	0
0	0	0	0	0	0	0	0	0	0
0	0	0	0	0	0	0	0	0	0
0	0	0	0	0	0	0	0	0	0
0	0	0	0	0	0	0	0	0	0
0	0	0	0	0	0	0	0	0	0
0	0	0	0	0	0	0	0	0	0
0	0	0	0	0	0	0	0	0	0

Table A.10
Thermal model output—household 1271 (electricity)

SHUEC	DSHUEC	ACUEC	DACUEC	TEMP
25,006	618	0	0	28.34
40,597	1,100	0	0	24.64
11,312	475	0	0	40.69
7,734	466	0	0	49.00
2,157	408	29	18	61.94
608	188	1,119	422	67.98
161	66	3,272	1,061	71.76
2,038	338	267	95	62.48
6,519	472	0	0	52.10
8,349	456	0	0	47.07
15,917	546	0	0	34.25
17,547	513	0	0	28.03
20,102	513	0	0	21.75
19,259	622	0	0	32.06
0	0	0	0	0

Table A.11
Thermal model output—household 1271 (natural gas)

SHUEC	DSHUEC	ACUEC	DACUEC	TEMP
11,476	518	0	0	42.63
7,081	785	344	105	58.71
886	272	4,228	1,311	69.57
10,218	865	246	67	55.36
26,947	1,020	0	0	37.58
64,208	2,073	0	0	32.05
0	0	0	0	0
0	0	0	0	0
0	0	0	0	0
0	0	0	0	0
0	0	0	0	0
0	0	0	0	0
0	0	0	0	0

while delivered energy per hour (Btuh) on a summer day is:

$$Q = 16798. + 758.09 \, (t-\tau) + 3.0979 \, (t-\tau)^2$$

These estimated relationships have not been adjusted for delivery losses or system coefficient of performance. Over the range in which the thermal mode is utilized, the relationships are quite linear. Note, however, that these functions embody the attributes of a particular structure with given insulation levels and may well shift remarkably from household to household.

Table A.14 presents the operating and capital costs for ten alternative HVAC systems facing household 1271 in 1962, the year of house construction. Costs have been normalized to 1967 dollars. The variables ACHEAT, SHEATN, SHEATD, and SHEATP are the predicted HVAC system capacities for air conditioning and for noncentral, duct, and hydronic space heating respectively. Details on capacity estimation and allocation of capital costs are given in Chapter 2 and in Cowing, Dubin, and McFadden (1982). A simple comparison of operating and capital costs indicates that household 1271 chooses gas hydronic system 3, which is dominated by other systems. The challenge of the discrete choice model is to describe the choice process adequately in the presence of unobserved cost components.

Table A.12
Selected variables from NIECS for household 1271

HDD4170	6848.	heating degree-days
CDD4170	387.	cooling degree-days
HDD7879	7057.	heating degree-days
CDD78	378.	cooling degree-days
NXELYR[a]	495.	dollars
NCELYRP[a]	10214.	kwh
NXNGYR[a]	567.	dollars
NCNGYRB[a]	1370.10	therms
ACUEC	2,112.	MBtu
DACUEC	248.	MBtu
SHUECE	151,340.	MBtu
DSHUECE	5,621.	MBtu
SHUECG	206,210.	MBtu
DSHUECG	7,659.	MBtu
WMPE78	.045172	dollars/kwh
SMPE78	.049483	dollars/kwh
OSMPE78	.045172	dollars/kwh
AEMPE78	.045172	dollars/kwh
AVEP78	.053329	dollars/kwh
AVGP78	.40778	dollars/therm
MPG78	.32767	dollars/therm

[a]Source: NIECS annual file.

Table A.13
Aggregated monthly billing data—household 1271

	electricity	natural gas
DAYS	363	362
QUAN (kwh/therms)	10,395	1,343
EXPEN	513	560
HDD65	6,766	6,586
CDD65	384	378
HDD75	10,150	9,964
CDD75	18	16
SHUECE	112,939	109,340
DSHUECE	5,028	5,015
ACUEC[a]	4,687	4,818
DACUEC	1,596	1,483

[a]ACUEC and DACUEC have not been adjusted to reflect system coefficient of performance. The coefficient of performance for air conditioning is 3.92 for household 1271.

Table A.14

Operating and capital costs of alternative HVAC
systems in year house built—household 1271
1967 dollars

OPCST1	497.39	CAPCST1	1226.80
OPCST2	519.88	CAPCST2	2345.60
OPCST3	490.84	CAPCST3	2821.40
OPCST7	203.63	CAPCST7	1834.40
OPCST8	226.12	CAPCST8	2694.00
OPCST9	200.95	CAPCST9	3304.60
OPCST13	1752.30	CAPCST13	982.61
OPCST14	1774.80	CAPCST14	2500.70
OPCST15	877.85	CAPCST15	7879.80
OPCST18	1611.90	CAPCST18	1154.60
ACHEAT	36.837	MBtuh	
SHEATN	59.348	MBtuh	
SHEATD	64.516	MBtuh	
SHEATP	63.666	MBtuh	

Conditional moments in the generalized extreme value family[1]

B.1. Introduction

The discrete/continuous econometric systems derived in Chapter 1 have the schematic form:

$$x = Z\beta^i + \eta \tag{1}$$

where Z is a vector of attributes of the alternatives and of the decision-maker, β^i is a vector of parameters that depend on the portfolio of appliances i, and η is an unobserved random variable whose density $f(\eta|i)$ depends in general on the observed choice i. The choice probabilities are functions of Z, the parameters β^i, and other parameters α:

$$P_i = \text{Prob(Portfolio } i \text{ is chosen)} = G^i(Z, \beta^1, \beta^2, ..., \beta^m, \alpha) \tag{2}$$

The likelihood of an observation (Z, i, x) is then:

$$G^i(Z, \beta^1, \beta^2, ..., \beta^m, \alpha) \cdot f(x - Z\beta^i|i) \tag{3}$$

Under standard regularity conditions, full information maximum likelihood (FIML) estimation of the parameters $\beta^1, ..., \beta^m, \alpha$ will yield consistent asymptotically efficient estimates, while maximum likelihood estimation of the discrete-choice model alone will yield consistent, but not usually efficient, estimates of the identifiable parameters.

[1]In the course of the exposition, several theorems related to the independent form of the generalized extreme value family, i.e., the multinomial logit model, are derived. Specifically, Corollary B.2.2 and Theorems B.3.3, B.4.1, B.4.2 and, B.4.3, which present the conditional moments for the multinomial logit model, are stated in Dubin and McFadden (1984). It should be further noted that Theorems B.3.3, B.4.1, and B.4.2 have been independently demonstrated by Hay (1980).

The system (1)-(3) is a variant of the "hybrid model with structural shift" analyzed in detail by Heckman (1978), and the estimators and properties he develops can be applied with straightforward modification. This system can also be interpreted as a "switching regression" with the structure analyzed by Lee (1981), Goldfeld and Quandt (1973, 1976), and Maddala and Nelson (1974). Note that (1) may be written:

$$x = Z\beta^i + E(\eta|i) + v \tag{4}$$

$$= \sum_{j=1}^{m} Z \delta_{ij} \beta^j + E(\eta|i) + v \tag{5}$$

$$= \sum_{j=1}^{m} Z P_j \beta^j + \xi \tag{6}$$

where:

$$\delta_{ij} = 1 \text{ iff } i=j , \quad E(\eta|i) = \int_{-\infty}^{+\infty} f(\eta|i) \, \eta \, d\eta$$

$$v = \eta - E(\eta|i) , \quad \xi = \eta + Z \sum_{j=1}^{m} (\delta_{ij} - P_j) \beta^j$$

Then $E(\eta|i) = 0$ and $E(\xi' Z P_j) = 0$ for $j = 1, 2, ..., m$. The choice probabilities P_i given by (2) and the conditional expectation $E(\eta|i)$ are nonlinear functions of the parameters of the problem. Under specific distributional assumptions, these functions may have computationally tractable forms. For example, if the discrete choice is binary and determined by a latent variable whose joint distribution with η is bivariate normal, then P_1 is a probit function and $E(\eta|i)$ is proportional to a Mill's ratio evaluated at the mean of the latent variables; see Heckman (1978).

Alternatively, suppose the utilities of different portfolios have independent extreme value distributions with a common variance. The choice probabilities are then multinomial logit; see McFadden (1973). The conditional expectation $E(\eta|i)$ is a simple function of the choice probabilities and their logs; see Dubin and McFadden (1984).

One method of estimating the parameters of (1) that is consistent under standard regularity conditions is to apply nonlinear least squares to (4) or (6). A second method is to replace $E(\eta|i)$ in (4) with a consistent estimate obtained by first estimating the parameters of the choice probabilities. Heckman (1979) shows that this procedure is also consistent under standard regularity conditions, although the asymptotic covariance matrices of the least squares coefficients obtained by this method differ from the standard formula due to the presence of estimated explanatory variables.[2]

The purpose of this appendix is to establish basic results on the conditional moments of generalized extreme value (GEV) random variables. This provides a useful generalization of the Heckman probit and the Dubin-McFadden logit selectivity corrections by including a class of dependent multinomial probability models. We introduce the GEV distribution and discuss its properties. We then derive the first, second, and cross-conditional moments for GEV variables given that a specific alternative has been selected. Finally, we allow the random variable η in equation (1) to have a linear conditional expectation in the space of GEV random variables and derive its properties. These results provide the distributional framework for consistent two-step estimation techniques.

B.2. Conditional moments in GEV

The following theorem due to McFadden (1978) introduces a general family of GEV choice models.

Theorem B.2.1. [McFadden]. Suppose $G(y_1, y_2, ..., y_J)$ is a nonnegative, homogeneous of degree one function of $(y_1, y_2, ..., y_J) \geqslant 0$. Suppose $\lim_{y_i \to +\infty} G(y_1, y_2, ..., y_J) = +\infty$ for $i = 1, 2, ..., J$. Suppose for any distinct $(i_1, i_2, ..., i_k)$ from $\{1, 2, ..., J\}$, $\partial^k G/\partial y_{i_1}, ..., \partial y_{i_k}$ is nonnegative if k is odd and nonpositive if k is even. Then:

$$P_i = e^{V_i} G_i(e^{V_1}, ..., e^{V_J})/G(e^{V_1}, ..., e^{V_J}) \tag{7}$$

---

[2] A third estimation technique applies the method of instrumental variables to (1) using consistent estimates of the choice probabilities as instruments. Consistency and efficiency of the various procedures are considered in Chapter 5.

defines a choice model that is consistent with random utility maximization.

Proof. Theorem B.2.1 is proved in two steps. The first step demonstrates that the function:

$$F(\varepsilon_1, \varepsilon_2, ..., \varepsilon_J) = \exp[- G(e^{-\varepsilon_1}, e^{-\varepsilon_2}, ..., e^{-\varepsilon_J})] \tag{8}$$

is a multivariate extreme value distribution. The details may be found in McFadden (1978).

The second step assumes a population of individuals with utilities $u_i = V_i + \varepsilon_i$, where $(\varepsilon_1, \varepsilon_2, ..., \varepsilon_J)$ is distributed F. Let ε denote the vector $(\varepsilon_1, \varepsilon_2, ..., \varepsilon_J)$. Then:

$$P_i = \text{Prob}[u_i \geqslant u_j, \forall i \mid i \neq j]$$

$$= \text{Prob}[V_i + \varepsilon_i \geqslant V_j + \varepsilon_j, \forall i \mid i \neq j] \tag{9}$$

may be written:

$$\int_{\varepsilon_i=-\infty}^{+\infty} F_i(<V_i + \varepsilon_i - V_j>) \, d\varepsilon_i \tag{10}$$

where F_i denotes the derivative of F with respect to its ith argument, and $<a_j>$ denotes a vector with jth component a_j. From (10) and the definition of the GEV distribution, we have:

$$P_i = \int_{-\infty}^{+\infty} \exp\left[-G[<e^{-(V_i+\varepsilon_i-V_j)}>] \right] G_i[<e^{-(V_i+\varepsilon_i-V_j)}>] \, e^{-\varepsilon_i} \, d\varepsilon_i$$

$$= \int_{-\infty}^{+\infty} e^{-\varepsilon_i} G_i[<e^{V_j}>] \exp\left[-G[<e^{V_j}>] \cdot e^{-V_i} e^{-\varepsilon_i} \right] \, d\varepsilon_i$$

$$= \frac{G_i[<e^{V_j}>]}{G[<e^{V_j}>]} \, e^{V_i} \quad \text{Q.E.D.} \tag{11}$$

The second equality in equation (11) uses the fact that G is homogeneous of degree one and the implication that G_i is homogeneous of degree zero. The third equality makes use of the result:

$$\int_{-\infty}^{+\infty} e^{-\varepsilon} \exp[-\theta^{-1} e^{-\varepsilon}] \, d\varepsilon = \theta^{-1} \tag{12}$$

which follows from the substitution $u \rightarrow -\theta e^{-\varepsilon}$.

Corollary B.2.1. [Multinomial Logit Model]. Let:

$$G[y] = [\sum_{j=1}^{J} y_j^{1/\theta}]^{\theta}. \quad \text{Then:} \quad P_i = \frac{e^{V_i/\theta}}{\sum_{j=1}^{J} e^{V_j/\theta}}$$

Proof. This result is found in McFadden (1978). One need simply verify the linear homogeneity of G and apply equation (1). Q.E.D.

McFadden shows that if $\varepsilon_j \rightarrow +\infty$ $\forall j \mid j \neq i$, then equation (2) implies, $F[\varepsilon_i] = \exp[-a_i e^{-\varepsilon}]$, where $a_i = G[0, ..., 0, 1, 0, ..., 0]$; with one in the ith coordinate. Under the assumptions of Corollary B.2.1, the marginal distribution, $F[\varepsilon_i]$, is $\exp[-e^{-\varepsilon_i}]$ (note $a_i = 1$); the cumulative distribution for an extreme value distributed random variable with variance $\pi^2/6$.

McFadden's proof of Theorem B.2.1 may be modified to demonstrate that:

$$F[\varepsilon_1, \varepsilon_2, ..., \varepsilon_J] = \exp\left[-G[<e^{-\varepsilon_i/\theta}>]\right] \tag{13}$$

is a multivariate extreme value distribution. Application of (4) implies the probabilistic choice system:

$$P_i = e^{V_i/\theta} G_i[<e^{V_j/\theta}>] / G[<e^{V_j/\theta}>] \tag{14}$$

In this case, the marginal distribution for ε_i becomes $\exp[-a_i e^{-\varepsilon_i/\theta}]$, which is the cumulative distribution function for an extreme value distributed random variable with variance $(\pi^2/6)\,\theta^2$.[3] When:

$$G[<y_j>] = \sum_{j=1}^{J} y_j$$

equation (14) implies choice probabilities of the multinomial logit form. Furthermore, ε_i has mean $\gamma\theta$ and variance $(1/6)\,\pi^2\theta^2$. More generally, ε_i will have mean u and variance $(1/6)\,\pi^2\theta^2$ when:

$$G[y_1, y_2, ..., y_J] = \frac{\exp(u/\theta)}{\exp(\gamma)} \cdot \left(\sum_{j=1}^{J} y_j\right)$$

Let $\delta_j(\varepsilon)$ be an indicator random variable that is 1 when j is the chosen alternative, i.e., when $V_j + \varepsilon_j \geq V_i + \varepsilon_i$, $\forall i \mid i \neq j$, and 0 otherwise. We write δ_j as a function of ε to emphasize that it is a random variable whose outcome, conditioned on the V_j 's, depends on the realization of ε. Lemma B.2.1 presents the conditional moments.[4]

Lemma B.2.1. Let ε be GEV distributed with cumulative distribution function $F(\varepsilon)$ given by equation (13). Let $g(\cdot)$ be an arbitrary real-valued function. Then:

(a) $E(g(\varepsilon_1)|\delta_1(\varepsilon) = 1) = E(g(\varepsilon)|\varepsilon \sim EV[\theta\,(\ln G_1 - \ln P_1), \theta])$

where $EV[a,b]$ denotes an extreme-valued distributed random variable with location parameter a and scale parameter b.

(b) Let G be additively separable with $G(y) = G^A(y^A) + y_2$ where $y = (y^A, y_2)$ and where $G^A(\cdot)$ is homogeneous of degree one. Let ε have the corresponding partition, i.e., $\varepsilon = (\varepsilon^A, \varepsilon_2)$. Then:

[3]The random variable ε_i has the properties: $E[\varepsilon_i] = \theta(\gamma + \ln a_i)$ and $Var[\varepsilon_i] = (1/6)\,\pi^2\theta^2$, where $\gamma = .5772156649 \cdots$ is Euler's constant.
[4]Note that without loss of generality, it suffices to consider the expressions $E(\varepsilon_1|\delta_1 = 1)$ and $E(\varepsilon_2|\delta_1 = 1)$ rather than the more general expression $E(\varepsilon_i|\delta_j = 1)$ for $i = j$ and for $i \neq j$.

$$E(\varepsilon_2 | \delta_1(\varepsilon) = 1)$$

$$= \frac{G[<e^{V_j/\theta}>]}{G^A[<e^{V_j/\theta}>]} \left[E\left[g(\varepsilon_2)|\varepsilon_2 \sim EV[0,\theta]\right] \right.$$

$$\left. - P_2 E\left[g(\varepsilon_2)|\varepsilon_2 \sim EV[-\theta (\ln P_2), \theta]\right] \right]$$

Proof.

(a) We make use of the properties of conditional expectations. Recall:

$$\int_{-\infty}^{y} \int_{x \in A} f(x, y) \, dx \, dy$$

$$= \text{Prob}[x \in A, Y \leqslant y]$$

$$= \text{Prob}[Y \leqslant y | x \in A] \, \text{Prob}[x \in A] \qquad (15)$$

Thus:

$$\text{Prob}[x \in A]^{-1} \int_{x \in A} f(x, y)dx = f(y | x \in A) \qquad (16)$$

Equation (16) implies:

$$E(Y | x \in A) = \int_y y \, f(y | x \in A) \, dy$$

$$= \text{Prob}[x \in A]^{-1} \int_y \int_{x \in A} y \, f(x, y) \, dx \, dy \qquad (17)$$

As an application of (11) we find:

$$E(g(\varepsilon_1)|\delta_1(\varepsilon) = 1)$$

$$= \frac{1}{P_1} \int_{\varepsilon_1=-\infty}^{\infty} \int_{\varepsilon_2=-\infty}^{V_1-V_2+\varepsilon_1} \cdots \int_{\varepsilon_J=-\infty}^{V_1-V_J+\varepsilon_1} g(\varepsilon_1)\, dF(\varepsilon)$$

$$= \frac{1}{P_1} \int_{\varepsilon=-\infty}^{\infty} g(\varepsilon)\, F_1[<\varepsilon + V_1 - V_j>]\, d\varepsilon$$

$$= \frac{1}{P_1} \int_{\varepsilon=-\infty}^{\infty} g(\varepsilon)\, e^{-\varepsilon/\theta}\, G_1\left[< e^{-(\varepsilon+V_1-V_j)/\theta} > \right] \cdot$$

$$\exp\left[-G[< e^{-(\varepsilon+V_1-V_j)/\theta} >] \right] \frac{d\varepsilon}{\theta}$$

$$= \frac{1}{P_1} \int_{\varepsilon=-\infty}^{\infty} g(\varepsilon)\, e^{-\varepsilon/\theta}\, G_1\left[<e^{V_j/\theta}> \right] \cdot$$

$$\exp\left(-G[<e^{V_j/\theta}>]\, e^{-\varepsilon/\theta}\, e^{-V_1/\theta} \right) \frac{d\varepsilon}{\theta}$$

Let $\theta_1 = G[<e^{V_j/\theta}>]\, e^{-V_1/\theta}$ and $\theta_2 = G_1[<e^{V_j/\theta}>]$. Then:

$$E(g(\varepsilon_1)|\delta_1(\varepsilon) = 1)$$

$$= \frac{\theta_2}{P_1} \cdot \int_{-\infty}^{\infty} g(\varepsilon)\, e^{-\varepsilon/\theta}\, \exp[-\theta_1\, e^{-\varepsilon/\theta}]\, \frac{d\varepsilon}{\theta}$$

$$= \frac{\theta_2}{P_1\, \theta_1} \cdot \int_{-\infty}^{\infty} g(\varepsilon)\, e^{-(\varepsilon-\theta\, k_1)/\theta}\, \exp[-e^{-(\varepsilon-\theta\, k_1)/\theta}]\, \frac{d\varepsilon}{\theta}$$

$$= E(g(\varepsilon)|\varepsilon \sim EV[\theta \ln \theta_1,\, \theta]) \tag{18}$$

where $k_1 = \ln \theta_1$ and $EV[a,b]$ denotes an extreme value distributed random variable with location parameter a and scale parameter b, i.e., $F_\varepsilon[t] = \exp[-e^{-(t-a)/b}]$.

From equation (14), $\theta_2/\theta_1 = G_1/\theta_1 = P_1$. Hence $\ln \theta_1 = (\ln G_1 - \ln P_1)$, so that substitution in the final equality of (18) proves the claim.

(b) $E(g(\varepsilon_2)|\delta_1(\varepsilon) = 1)$

$$= \frac{1}{P_1} \int_{\varepsilon_1=-\infty}^{+\infty} \int_{\varepsilon_2=-\infty}^{V_1-V_2+\varepsilon_1} \cdots \int_{\varepsilon_J=-\infty}^{V_1-V_J+\varepsilon_1} g(\varepsilon_2) \, dF(\varepsilon)$$

$$= \frac{1}{P_1} \int_{\varepsilon_1=-\infty}^{+\infty} \int_{\varepsilon_2=-\infty}^{V_1-V_2+\varepsilon_1} g(\varepsilon_2) \cdot$$

$$F_{12}[\varepsilon_1, \varepsilon_2, V_1-V_3+\varepsilon_1, \ldots, V_1-V_J+\varepsilon_1] \, d\varepsilon_2 \, d\varepsilon_1$$

$$= \frac{1}{P_1} \int_{\varepsilon_2=-\infty}^{+\infty} \int_{\varepsilon_2+V_2-V_1}^{+\infty} g(\varepsilon_2) \cdot$$

$$F_{12}[\varepsilon_1, \varepsilon_2, V_1-V_3+\varepsilon_1, \ldots, V_1-V_J+\varepsilon_1] \, d\varepsilon_1 \, d\varepsilon_2 \qquad (19)$$

From equation (13):

$$F(\varepsilon) = \exp\left[- G[<e^{-\varepsilon_i/\theta}>] \right] \qquad (20)$$

$$= \exp\left[- G^A[<e^{-\varepsilon_i'/\theta}>] \right] \cdot \exp[-e^{-\varepsilon_2/\theta}]$$

so that:

$$F_{12}(\varepsilon) = \exp\left[- G^A[<e^{-\varepsilon_i'/\theta}>] \right] G_1^A[<e^{-\varepsilon_i'/\theta}>] \, e^{-\varepsilon_1/\theta} \frac{1}{\theta} \qquad (21)$$

$$\cdot \exp[-e^{-\varepsilon_2/\theta}]\, e^{-\varepsilon_2/\theta}\, \frac{1}{\theta}$$

Hence:

$$F_{12}(\varepsilon_1,\, \varepsilon_2,\, V_1-V_3+\varepsilon_1,\, ...,\, V_1-V_J+\varepsilon_1)$$

$$= \exp[\,-\,G^A[e^{-\varepsilon_1/\theta},\, <e^{\frac{-V_1+V_j-\varepsilon_1}{\theta}}>]\,]\cdot$$

$$G_1^A\,[\,e^{-\varepsilon_1/\theta},<e^{\frac{-V_1+V_j-\varepsilon_1}{\theta}}>\,]\,e^{-\varepsilon_1/\theta}\,\frac{1}{\theta}\,\exp\!\left[-e^{-\varepsilon_2/\theta}\right]\,e^{-\varepsilon_2/\theta}\,\frac{1}{\theta}$$

$$= \exp\!\left[-e^{-\varepsilon_1/\theta}e^{-V_1/\theta}\cdot G^A[<e^{V_j/\theta}>]\right]\cdot$$

$$G_1^A[<e^{V_j/\theta}>]\,e^{-\varepsilon_1/\theta}\,\frac{1}{\theta}\cdot\exp[-e^{-\varepsilon_2/\theta}]\,e^{-\varepsilon_2/\theta}\,\frac{1}{\theta}\quad\text{Thus}\qquad(22)$$

$$E(g(\varepsilon_2)|\delta_1(\varepsilon)=1)$$

$$= \frac{G_1^A[<e^{V_j/\theta}>]}{P_1}\cdot\int_{\varepsilon_2=-\infty}^{\infty}g(\varepsilon_2)\,e^{-\varepsilon_2/\theta}\,\exp[-e^{-\varepsilon_2/\theta}]\cdot$$

$$\int_{\varepsilon_2+V_2-V_1}^{\infty}\exp\!\left[-e^{-\varepsilon_1/\theta}\,e^{-V_1/\theta}\,G^A[<e^{V_j/\theta}>]\right]\,e^{-\varepsilon_1/\theta}\,\frac{d\varepsilon_1}{\theta}\,\frac{d\varepsilon_2}{\theta}$$

$$= \frac{G_1^A[<e^{V_j/\theta}>]}{P_1\cdot\theta_1^A}\int_{\varepsilon_2=-\infty}^{\infty}g(\varepsilon_2)\,e^{-\varepsilon_2/\theta}\,\exp[-e^{-\varepsilon_2/\theta}]\cdot$$

$$\left[1-\exp\!\left(-\theta_1^A\,e^{\frac{-\varepsilon_2-V_2+V_1}{\theta}}\right)\right]\,\frac{d\varepsilon_2}{\theta}$$

where $\theta_1^A = e^{-V_1/\theta} \, G^A[<e^{V_j/\theta}>]$. Thus:

$$E(g(\varepsilon_2)|\delta_1(\varepsilon) = 1)$$

$$= \frac{G_1^A[<e^{V_j/\theta}>]}{P_1 \cdot \theta_1^A} \, E(g(\varepsilon_2)|\varepsilon_2 \sim EV[0,\theta])$$

$$- \frac{G_1^A[<e^{V_j/\theta}>]}{P_1 \cdot \theta_1^A} \cdot \int_{\varepsilon_2=-\infty}^{\infty} g(\varepsilon_2) \, e^{-\varepsilon_2/\theta} \exp[-e^{-\varepsilon_2/\theta}] \cdot$$

$$\exp[-\theta_1^A \, e^{\frac{-\varepsilon_2 - V_2 + V_1}{\theta}}] \, \frac{d\varepsilon_2}{\theta} \tag{23}$$

But:

$$\int_{\varepsilon_2=-\infty}^{\infty} g(\varepsilon_2) \, e^{-\varepsilon_2/\theta} \exp[-e^{-\varepsilon_2/\theta} \cdot \theta_2] \, \frac{d\varepsilon_2}{\theta}$$

$$= \frac{1}{\theta_2} \, E\left[g(\varepsilon_2)|\varepsilon_2 \sim EV[\theta \ln \theta_2, \theta] \right]$$

where we define $\theta_2 = (1 + e^{(V_1-V_2)/\theta} \cdot \theta_1^A)$. Hence:

$$E(g(\varepsilon_2)|\delta_1(\varepsilon) = 1)$$

$$= \frac{G_1^A[<e^{V_j/\theta}>]}{P_1 \cdot \theta_1^A} \cdot \left[E\left[g(\varepsilon_2)|\varepsilon_2 \sim EV[0, \theta] \right] \right.$$

$$\left. - \frac{1}{\theta_1} \cdot E\left[g(\varepsilon_2)|\varepsilon_2 \sim EV[\theta \ln \theta_2, \theta] \right] \right] \tag{24}$$

Note that $G_1^A[<e^{V_j/\theta}>] = G_1[<e^{V_j/\theta}>]$ implies:

$$\frac{G_1^A[<e^{V_j/\theta}>]}{P_1 \cdot \theta_1^A} = \frac{G_1[<e^{V_j/\theta}>] \, e^{V_1/\theta}}{G^A[<e^{V_j/\theta}>] \, P_1} = \frac{G[<e^{V_j/\theta}>]}{G^A[<e^{V_j/\theta}>]} \tag{25}$$

Also:

$$\theta_2 = \left(1 + e^{(V_1-V_2)/\theta}\right) \theta_1^A = \left(1 + e^{-V_2/\theta} \, G^A[<e^{V_j/\theta}>]\right)$$

$$= e^{-V_2/\theta} \left[e^{V_2/\theta} + G^A[<e^{V_j\theta}>]\right] = e^{-V_2/\theta} \, G[<e^{V_j/\theta}>] \tag{26}$$

Furthermore, $G_2/\theta_2 = P_2$ and $G_2 \equiv 1$ imply:

$$\frac{e^{V_2/\theta}}{G[<e^{V_j/\theta}>]} = \frac{1}{\theta_2} = P_2 \tag{27}$$

Combining equations (25) and (27) with equation (24), we find:

$$E(g(\varepsilon_2)|\delta_1(\varepsilon) = 1)$$

$$= \frac{G[<e^{V_j/\theta}>]}{G^A[<e^{V_j/\theta}>]} \cdot \left[E\left[g(\varepsilon_2)|\varepsilon_2 \sim EV[0, \theta]\right]\right.$$

$$\left. - P_2 \cdot E\left[g(\varepsilon_2)| \, \varepsilon_2 \sim EV[\theta \ln \theta_2, \theta]\right]\right] \tag{28}$$

From equation (27) we have:

$$\ln \theta_2 = \ln G_2 - \ln P_2 = -\ln P_2 \tag{29}$$

Combining (28) and (29) with (27) proves the claim.   Q.E.D.

As an application of Lemma B.2.1 we have:

Theorem B.2.2. Let ε be GEV distributed with cumulative distribution function $F(\varepsilon)$ given in (13). Then:

(a) $E(\varepsilon_1|\delta_1(\varepsilon) = 1) = \theta\,(\gamma + \ln G_1 - \ln P_1)$

(b) $E(\varepsilon_1^2|\delta_1(\varepsilon) = 1) = (\pi^2/6)\,\theta^2 + \theta^2\,(\gamma + \ln G_1 - \ln P_1)^2$

Let G be additively separable with $G(y) = G^A(y^A) + y_2$ where $y = (y^A, y_2)$ and where $G^A(\cdot)$ is homogeneous of degree one. Let ε have the corresponding partition, i.e., $\varepsilon = (\varepsilon^A, \varepsilon_2)$. Then:

(c) $E(\varepsilon_2|\delta_1(\varepsilon) = 1) = \dfrac{G[<e^{V_j/\theta}>]}{G^A[<e^{V_j/\theta}>]} \cdot \theta \cdot \left[(1-P_2)\,\gamma + P_2 \ln P_2 \right]$

(d) $E(\varepsilon_2^2|\delta_1(\varepsilon) = 1) = \dfrac{G[<e^{V_j/\theta}>]}{G^A[<e^{V_j/\theta}>]} \cdot \theta^2 \cdot$

$$\left[\gamma^2 - P_2\,(\gamma - \ln P_2)^2 + (1-P_2)\,(\pi^2/6) \right]$$

Proof.

(a) Using Lemma B.2.1 (a) with $g(\varepsilon) = \varepsilon$, we have:

$$E(\varepsilon_1|\delta_1(\varepsilon) = 1) = E(\varepsilon|\varepsilon \sim EV[\theta\,(\ln G_1 - \ln P_1),\,\theta])$$

$$= \theta\,(\gamma + \ln G_1 - \ln P_1) \tag{30}$$

(b) We take $g(\varepsilon) = \varepsilon^2$ so that:

$$E(\varepsilon_1^2 | \delta_1(\varepsilon) = 1)$$

$$= E\left[\varepsilon^2 | \varepsilon \sim EV[\theta \, (\ln G_1 - \ln P_1), \, \theta]\right]$$

$$= \left[E(\varepsilon | \varepsilon \sim EV[\theta \, (\ln G_1 - \ln P_1), \, \theta])\right]^2$$

$$+ \, var \left[\varepsilon | \varepsilon \sim EV[\theta \, (\ln G_1 - \ln P_1), \, \theta]\right]$$

$$= \theta^2 \, [\gamma + \ln G_1 - \ln P_1]^2 + (\pi^2/6) \, \theta^2 \qquad\qquad (31)$$

(c) Using Lemma B.2.1 (b) with $g(\varepsilon) = \varepsilon$, we have:

$$E(\varepsilon_2 | \delta_1(\varepsilon) = 1)$$

$$= (\frac{G}{G^A}) \cdot \left[E\left[\varepsilon_2 | \varepsilon_2 \sim EV[0, \, \theta]\right] - \right.$$

$$\left. P_2 \, E\left[\varepsilon_2 | \varepsilon_2 \sim EV[-\theta \ln P_2, \, \theta]\right] \right]$$

$$= (\frac{G}{G^A}) \cdot \left[\gamma\theta - P_2 \, (\gamma\theta - \theta \ln P_2) \right]$$

$$= (\frac{G}{G^A} \cdot \theta) \cdot \left[(1 - P_2) \, \gamma + P_2 \ln P_2 \right] \qquad\qquad (32)$$

(d) Using Lemma B.2.1 (b) with $g(\varepsilon) = \varepsilon^2$, we have:

$$E(\varepsilon_2^2 | \delta_1(\varepsilon) = 1)$$

$$= (\frac{G}{G^A}) \cdot \left[E \left[\varepsilon_2^2 | \varepsilon_2 \sim EV[0, \theta]\right] \right.$$

$$\left. - P_2 \cdot E \left[\varepsilon_2^2 | \varepsilon_2 \sim EV[-\theta \ln P_2, \theta]\right] \right]$$

$$= (\frac{G}{G^A}) \cdot \left[\left[(\gamma\theta)^2 + (\pi^2/6)\,\theta^2\right] - P_2 \left[\theta^2\,(\gamma - \ln P_2 + (\pi^2/6)\,\theta^2\right] \right]$$

$$= (\frac{G}{G^A}) \cdot \left[(\gamma\theta)^2 - P_2\,\theta^2\,(\gamma - \ln P_2)^2 + (1-P_2)\,(\pi^2/6)\,\theta^2 \right] \tag{33}$$

$$= (\frac{G}{G^A} \cdot \theta^2) \cdot \left[\gamma^2 - P_2\,(\gamma - \ln P_2)^2 + (1-P_2)\,(\pi^2/6) \right] \quad \text{Q.E.D.}$$

Comments: Theorem B.2.2 imposes strong separability in the functional form for G to obtain a closed form conditional expectation. When G has the additive from $G[y] = G^A[y_2^A] + y_2$, ε_2 is independent from ε^A. If we do not impose strong separability, then $F_{12}(\varepsilon)$ becomes:

$$F_{12}(\varepsilon) = \exp\left[-G[<e^{-\varepsilon_j/\theta}>] \right] e^{-\varepsilon_1/\theta} e^{-\varepsilon_2/\theta} \frac{1}{\theta^2} \cdot$$

$$\left[G_1[<e^{-\varepsilon_j/\theta}>]\, G_2[<e^{-\varepsilon_j/\theta}>] - G_{12}[<e^{-\varepsilon_j/\theta}>] \right] \tag{34}$$

Following the proof of Lemma B.2.1 (b), we see that the analogue of (22) for equation (34) does not permit an easy integration in (17). It is possible, however, to extend the results of Theorems B.2.2 (c) and (d) for $G[y] = G^A[y^A] + \alpha y_2$. We present the results in Corollary B.2.2.

Corollary B.2.2 Let ε be GEV distributed with cumulative distribution function $F(\varepsilon)$ given in (13). Let G be additively separable with $G(y) = G^A(y^A) + \alpha y_2$ where $y = (Y^A, y_2)$ and where $G^A(\cdot)$ is homogeneous of degree one. Define $\alpha^* = \theta \ln \alpha$. Then:

(a) $E(\varepsilon_2|\delta_1(\varepsilon) = 1) = \dfrac{G[<e^{V_j/\theta}>]}{G^A[<e^{V_j/\theta}>]} \cdot \left[(\gamma\theta + \alpha^*)(1-P_2) + \theta\, P_2 \ln P_2 \right]$

(b) $E(\varepsilon_2^2|\delta_1(\varepsilon) = 1) = \dfrac{G[<e^{V_j/\theta}>]}{G^A[<e^{V_j/\theta}>]} \cdot$

$\left[(\pi^2/6)\, \theta^2\, (1 - P_2) + (\gamma\theta + \alpha^*)^2\, (1 - P_2) \right.$

$\left. + 2\,\theta\,(\gamma\theta + \alpha^*)\, P_2 \ln P_2 - \theta^2\, P_2(\ln P_2)^2 \right]$

Proof. The proof of Corollary B.2.2 requires minor modifications in the arguments that demonstrate Lemma B.2.1 (b) and Theorems B.2.2 (c) and (d). It is therefore omitted.   Q.E.D.

As an illustration of Theorem B.2.2 and its corollary, we derive the conditional moments for the multinomial and nested logit models.

Example B.2.1. [Conditional Moments in the Multinomial Logit Model].

Let: $G[y] = \alpha \, [\sum\limits_{j=1}^{J} y_j]$ and $\alpha^* = \theta \ln \alpha$. Then:

(a) $E(\varepsilon_1|\delta_1(\varepsilon) = 1) = (\alpha^* + \gamma\theta) - \theta \ln P_1$

(b) $E(\varepsilon_1^2|\delta_1(\varepsilon) = 1) = \pi^2\, \theta^2/6 + (\alpha^* + \gamma\theta)^2 + \theta^2\,(\ln P_1)^2$

$- 2\,(\alpha^* + \gamma\theta) \cdot \theta\,(\ln P_1)$

(c) $E(\varepsilon_2|\delta_1(\varepsilon) = 1) = (\alpha^* + \gamma\theta) + \theta\, P_2\, (\ln\, P_2)/(1-P_2)$

(d) $E(\varepsilon_2^2|\delta_1(\varepsilon) = 1) = \pi^2\theta^2/6 + (\alpha^* + \gamma\theta)^2$

$\quad - P_2\, \theta^2\, (\ln\, P_2)^2/(1-P_2) + 2\, (\alpha^* + \gamma\theta)\, (\theta\, \ln\, P_2)\, P_2/(1-P_2)$

Proof.

(a) $G_1 = \alpha$ and $\theta \ln G_1 = \theta \ln \alpha = \alpha^*$. Apply Theorem B.2.2 (a).

(b) Use Theorem B.2.2 (b) with $G_1 = \alpha$. Then:

$$E(\varepsilon_1^2|\delta_1(\varepsilon) = 1) = \frac{\pi^2}{6}\, \theta^2 + \theta^2\, (\gamma + \ln\, \alpha - \ln\, P_1)^2$$

$$= \frac{\pi^2}{6}\, \theta^2 + \theta^2\, (\gamma + \ln\, \alpha)^2 - 2\, \theta^2\, (\gamma + \ln\, \alpha)\, (\ln\, P_1) + \theta^2\, (\ln\, P_1)^2$$

$$= \frac{\pi^2}{6}\theta^2 + (\gamma\theta + \alpha^*)^2 - 2\, (\gamma\theta + \alpha^*)\, \theta\, (\ln\, P_1) + \theta^2\, (\ln\, P_1)^2$$

(c) Apply Corollary B.2.2 (a) with $G^A[y^A] = \alpha\, (\sum_{j\neq 2}\, y_j\,)$. Then:

$$E(\varepsilon_2|\delta_1(\varepsilon) = 1) = \frac{G[<e^{V_j/\theta}>]}{G^A[<e^{V_j/\theta}>]}\quad.$$

$$(\gamma\theta + \alpha^*)\, (1-P_2) + \theta\, P_2\, \ln\, P_2$$

From equation (14):

$$\frac{G[<e^{V_j/\theta}>]}{G^A[<e^{V_j/\theta}>]} = \frac{\alpha\, \sum\limits_{j-1}^{J}\, e^{V_j/\theta}}{\alpha\, \sum\limits_{j\neq 2}^{J}\, e^{V_j/\theta}} = 1/(1-P_2) \qquad (35)$$

Therefore:

$$E(\varepsilon_2|\delta_1(\varepsilon) = 1) = (\gamma\theta + \alpha^*) + \theta\, P_2\, (\ln P_2)/(1-P_2)$$

(d) Apply Corollary B.2.2 (b) with $G^A[y^A] = \alpha \sum\limits_{j \neq 2}^{J} y_j$ and equation (35). Then:

$$E(\varepsilon_2^2|\delta_1(\varepsilon) = 1) = \frac{\pi^2}{6}\,\theta^2 + (\gamma\theta + \alpha^*)^2 + 2\theta\,(\gamma\theta + \alpha^*) \cdot$$

$$P_2\, (\ln P_2)\,/\,(1-P_2) - \theta^2\, P_2\, (\ln P_2)^2\,/\,(1-P_2) \quad \text{Q.E.D.}$$

As a second illustration of Theorem B.2.2 and Corollary B.2.2, we derive the conditional moments for the nested logit model.

Example B.2.2. Consider a two-level nested logit model with three alternatives:

$$G[y_1, y_2, y_3] = \left[y_1^{1/(1-\sigma)} + y_3^{1/(1-\sigma)} \right]^{(1-\sigma)} + y_2 \tag{36}$$

Following McFadden (1978), one may verify that (36) satisfies the conditions of Theorem B.2.1. Therefore, application of equation (14) yields:

$$P[2|1, 2, 3] = \frac{e^{V_2/\theta}}{\left[e^{V_1/\theta(1-\sigma)} + e^{V_3/\theta(1-\sigma)} \right]^{(1-\sigma)} + e^{V_2/\theta}} \tag{37}$$

$$P[1|1, 2, 3] = \frac{\left[e^{V_1/\theta(1-\sigma)} + e^{V_3/\theta(1-\sigma)} \right]^{(1-\sigma)}}{\left[e^{V_1/\theta(1-\sigma)} + e^{V_3/\theta(1-\sigma)} \right]^{(1-\sigma)} + e^{V_2/\theta}} \cdot$$

$$\frac{e^{V_1/\theta(1-\sigma)}}{\left[e^{V_1/\theta(1-\sigma)} + e^{V_3/\theta(1-\sigma)} \right]} \tag{38}$$

$$= P[(1, 3)|(1, 2, 3)] \cdot P[1|(1, 3)]$$

where $P(i|A)$ denotes the probability that i is chosen from the set A. From equation (36) we find:

$$G_1 = \left[e^{V_1/\theta(1-\sigma)} + e^{V_3/\theta(1-\sigma)} \right]^{-\sigma} \cdot e^{\sigma V_1/\theta(1-\sigma)}$$

$$= P[1|1, 3]^{\sigma} \qquad (39)$$

We define $G^A[y_1, y_2, y_3] = [y_1^{1/(1-\sigma)} + y_3^{1/(1-\sigma)}]^{(1-\sigma)}$. It follows that:

$$\left(\frac{G}{G^A}\right)[<e^{V_j/\theta}>] = \frac{\left[e^{V_1/\theta(1-\sigma)} + e^{V_3/\theta(1-\sigma)} \right]^{(1-\sigma)}}{\left[e^{V_1/\theta(1-\sigma)} + e^{V_3/\theta(1-\sigma)} \right]^{(1-\sigma)}}$$

$$+ \frac{e^{V_2/\theta}}{\left[e^{V_1/\theta(1-\sigma)} + e^{V_3/\theta(1-\sigma)} \right]^{(1-\sigma)}}$$

$$= 1 + \frac{P[2|(1, 2, 3)]}{P[(1, 3)|(1, 2, 3)]}$$

$$= 1/(1 - P[2|(1, 2, 3)]) \qquad (40)$$

For G given in equation (36), Theorem B.2.2 implies:

$$E(\varepsilon_1|\delta_1(\varepsilon) = 1) = \theta \left[\gamma + \sigma \ln P(1|1, 3) - \ln P(1|1, 2, 3) \right] \tag{41}$$

$$E(\varepsilon_1^2|\delta_1(\varepsilon) = 1) = \frac{\pi^2}{6} \theta^2 + \theta^2 \left[\gamma + \sigma \ln P(1|1, 3) \right.$$

$$\left. - \ln P(1|1, 2, 3) \right]^2 \tag{42}$$

Application of Theorems B.2.2 (c) and (d) to equation (34) implies:[5]

$$E(\varepsilon_2|\delta_1(\varepsilon) = 1) = \theta \left[\gamma + P_2 (\ln P_2)/(1-P_2) \right] \tag{43}$$

and:

$$E(\varepsilon_2^2|\delta_1(\varepsilon) = 1) = \theta^2 \left[\frac{\pi^2}{6} + \left[\gamma^2 - P_2 (\gamma - \ln P_2)^2 \right]/(1-P_2) \right] \tag{44}$$

In equations (41) and (42), one observes that the nested logit model implies a closed-form expression in the conditional probabilities of reaching alternative one from various nodes of the tree. The conditional expectations in (41) and (42) differ from their counterparts derived in the multinomial logit example by the term $\sigma \ln P(1|(1, 3))$. As σ tends to zero in the limit, the nested logit model converges to the multinomial logit model and the term $\sigma \ln P(1|(1, 3))$ vanishes.

Comparison of (37), (38), and the corresponding expressions in the multinomial logit example reveals equal conditional expectations. The essence of the separability assumption is that the variable ε_2 behaves as a multinomial rather than nested logit random variable.

The calculations involved in equations (41)—(44) are easily modified to trees of any depth or size. As an illustration, consider a two-level nested logit model with M alternatives:

$$G(y) = \sum_{m=1}^{M} a_m \left[\sum_{i \in B_m} y_i^{1/(1-\sigma_m)} \right]^{1-\sigma_m} \tag{45}$$

[5]Expressions for the conditional expectation of ε_3 are analogous to those presented for ε_1.

where $B_m \subseteq \{1, 2, ..., J\}$, $\bigcup_{m=1}^{M} B_m = \{1, 2, ..., J\}$, $a_m > 0$, and $0 \leqslant \sigma_m < 1$. McFadden (1978) derives the choice probabilities for equation (45) and shows that they satisfy:

$$P_i = \frac{\displaystyle\sum_{m \mid i \in B_m}^{M} e^{V_i/(1-\sigma_m)} \, a_m \left[\sum_{j \in B_m} e^{V_j/(1-\sigma_m)} \right]^{-\sigma_m}}{\displaystyle\sum_{n-1}^{M} a_n \left[\sum_{k \in B_n} e^{V_k/(1-\sigma_n)} \right]^{(1-\sigma_n)}}$$

$$= \sum_{m \mid i \in B_m} P[i \mid B_m] \cdot P[B_m] \tag{46}$$

where:

$$P[i \mid B_m] = \begin{cases} e^{V_i/(1-\sigma_m)} / \displaystyle\sum_{j \in B_m} e^{V_j/(1-\sigma_m)} & \text{if } i \in B_m \\ 0 & \text{otherwise} \end{cases} \tag{47}$$

and:

$$P[B_m] = \frac{a_m \left[\displaystyle\sum_{j \in B_m} e^{V_j/(1-\sigma_m)} \right]^{(1-\sigma_m)}}{\displaystyle\sum_{n-1}^{M} a_n \left[\sum_{k \in B_n} e^{V_k/(1-\sigma_n)} \right]^{(1-\sigma_n)}} \tag{48}$$

From equation (45) we have:

$$G_i(y) = \sum_{m \mid i \in B_m}^{M} a_m \left[\sum_{j \in B_m} y_j^{1/(1-\sigma_m)} \right]^{-\sigma_m} \cdot y_i^{\sigma_m/(1-\sigma_m)} \tag{49}$$

so that:

$$G_i(<e^{V_j}>) = \sum_{m=1}^{M} a_m \, P[i \,|\, B_m]^{\sigma_m} \tag{50}$$

The form of the derivative in (50) generalizes in higher order trees. As an example consider the three-level tree structure:

$$G = \sum_a \left[\sum_d \left[\sum_m y_{mda}^{1/(1-\sigma)} \right]^{(1-\sigma)/(1-\delta)} \right]^{(1-\delta)} \tag{51}$$

In this case one may show:

$$G_{mda}[<e^{V_j}>] = \sum_a \sum_d P[d \,|\, a]^\delta \cdot P[m \,|\, da]^\sigma \tag{52}$$

where G_{mda} denotes the derivative of G in (51) with respect to y_{mda}. Further, equation (40) generalizes to cases in which G exhibits strong separability in some of its arguments. For example, suppose:

$$G = G^A + a_{M+1} \, y_{M+1} \, . \quad \text{Then:} \quad P_{M+1} = a_{M+1} \, e^{V_{M+1}/\theta}/G$$

and:

$$((G-G^A)/G)\,(<e^{V_j/\theta}>) = P_{M+1}$$

Thus:

$$(G/G^A)\,(<e^{V_j/\theta}>) = (1 - P_{M+1})^{-1}$$

as in equation (40).

B.3. Conditional covariance in the GEV family

We now consider the conditional moment of the product of two GEV random variables. Rather than calculate $E[\varepsilon_1\varepsilon_2|\delta_1(\varepsilon) = 1]$, we will alternatively find $E[(\varepsilon_2 - \varepsilon_1)^2|\delta_1(\varepsilon) = 1]$ and use the relation $(\varepsilon_2 - \varepsilon_1)^2 = \varepsilon_2^2 - 2\varepsilon_1\varepsilon_2 + \varepsilon_1^2$ in conjunction with Theorem B.2.2. It is well known that the difference $(\varepsilon_2 - \varepsilon_1)$ has a logistic distribution when ε_1 and ε_2 are independent identically extreme value distributed. Our next result finds the joint distribution function for $(Y_2, Y_3, ..., Y_J) = (\varepsilon_2 - \varepsilon_1, \varepsilon_3 - \varepsilon_1, ..., \varepsilon_J - \varepsilon_1)$ when ε has a GEV distribution.

Theorem B.3.1. [Generalized Logistic Distribution]. Let $Y_j = \varepsilon_j - \varepsilon_1$ for $j = 2, 3, ..., J$ where ε has a GEV distribution given by equation (13). Then:

$$H[w_2, w_3, ..., w_J] = Prob[Y_2 \leqslant w_2, Y_3 \leqslant w_3, ..., Y_J \leqslant w_J]$$

$$= G_1[<e^{-w_j/\theta}>]/G[<e^{-w_j/\theta}>]$$

with $w_1 \equiv 0$.

Proof.

$$H = Prob[Y_2 \leqslant w_2, ..., Y_J \leqslant w_J]$$

$$= \int_{\varepsilon_1=-\infty}^{\infty} \int_{\varepsilon_2=-\infty}^{\varepsilon_1+w_2} \cdots \int_{\varepsilon_J=-\infty}^{\varepsilon_1+w_J} dF(\varepsilon)$$

$$= \int_{\varepsilon_1=-\infty}^{\infty} F_1[\varepsilon, \varepsilon+w_2, ..., \varepsilon+w_J]\, d\varepsilon$$

$$= \int_{\varepsilon=-\infty}^{\infty} \exp\left[-G[<e^{-(\varepsilon+w_j)/\theta}>]\right]G_1[<e^{(-\varepsilon-w_j)/\theta}>]e^{-\varepsilon/\theta}\frac{d\varepsilon}{\theta}$$

$$= \int_{\varepsilon=-\infty}^{\infty} \exp\left[-e^{-\varepsilon/\theta}G[<e^{-w_j/\theta}>]\right]G_1[<e^{-w_j/\theta}>]e^{-\varepsilon/\theta}\frac{d\varepsilon}{\theta}$$

$$= \frac{G_1[<e^{-w_j/\theta}>]}{G[<e^{-w_j/\theta}>]} \quad \text{Q.E.D.}$$

Two familiar results follow immediately from Theorem B.3.1.

Corollary B.3.1.

(a) $H[V_1-V_2, V_1-V_3, ..., V_1-V_J] = P_1$

(b) $(Y_2, Y_3, ..., Y_J)$ is logistically distributed when $G[y] = \sum_{j-1}^{J} y_j$.

Proof.

(a) $H[V_1-V_2, ..., V_1-V_J] = \dfrac{G_1[<e^{-(V_1-V_j)/\theta}>]}{G[<e^{-(V_1-V_j)/\theta}>]}$

$$= e^{V_1/\theta} G_1[<e^{V_j/\theta}>] G[<e^{V_j/\theta}>] = P_1$$

where the first equality uses the result of Theorem B.3.1, the second equality uses the homogeneity property of G, and the third equality uses equation (14).

(b) Because $G[y] = \sum_{j-1}^{J} y_j$, $G_1[y] = 1$.

Theorem B.3.1 implies:

$$H[w_2, ..., w_J] = \left[\sum_{j-1}^{J} e^{-w_j/\theta} \right]^{-1}$$

which is a multivariate logistic distribution.   Q.E.D.

We now make the assumption that ε_1 and ε_2 are independent from each other and from ε^A.

Theorem B.3.2. Let ε be GEV distributed with $G[y] = \alpha\,y_1 + \alpha\,y_2 + \alpha\,G^A[<y_j^A>]$ where G^A is homogeneous of degree one and where $y = (y_1, y_2, y^A)$. Then:

$$E((\varepsilon_2 - \varepsilon_1)^2|\delta_1(\varepsilon) = 1)$$

$$= \theta^2\,[\ln\,((1-P_2)/P_1)]^2 - 2\theta^2\,\ln\,((1-P_2)/P_1)\,\cdot$$

$$\left[P_2\,\ln\,P_2/(1-P_2) + \ln\,(1-P_2)\right] + \theta^2/(1-P_2)\,\cdot$$

$$\int_{-\infty}^{\ln\,[(1-P_2)/P_2]} h(z)\,dz \quad \text{with} \quad h(z) = \frac{z^2\,e^{-z}}{(1+e^{-z})^2}$$

Proof.

$$E((\varepsilon_2 - \varepsilon_1)^2|\delta_1(\varepsilon) = 1)$$

$$= \frac{1}{P_1}\,\int_{\varepsilon_1=-\infty}^{\infty}\int_{\varepsilon_2=-\infty}^{V_1-V_2+\varepsilon_1} (\varepsilon_2 - \varepsilon_1)^2\,\cdot$$

$$F_{12}[\varepsilon_1,\,\varepsilon_2,\,V_1-V_3+\varepsilon_1,\,...,\,V_1-V_J+\varepsilon_1]\,d\varepsilon_2\,d\varepsilon_1$$

We make the logistic transformation: $z_1 = \varepsilon_1$, $z_2 = \varepsilon_2 - \varepsilon_1$. It is easily verified that this transformation has unit Jacobian. Thus:

$$E((\varepsilon_2 - \varepsilon_1)^2|\delta_1(\varepsilon) = 1)$$

$$= \frac{1}{P_1}\,\int_{z_1=-\infty}^{\infty}\int_{z_2=-\infty}^{V_1-V_2} z_2^2\,\cdot$$

$$F_{12}[z_1, z_1 + z_2, V_1 - V_3 + z_1, ..., V_1 - V_J + z_1] \, dz_2 \, dz_1$$

$$= \frac{1}{P_1} \int_{z_2=-\infty}^{V_1-V_2} z_2^2 \int_{z_1=-\infty}^{\infty}$$

$$F_{12}[z_1, z_1 + z_2, V_1 - V_3 + z_1, ..., V_1 - V_J + z_1] \, dz_1 \, dz_2$$

Define:

$$H[w_2, ..., w_J] = \int_{\varepsilon=-\infty}^{\infty} F_1[\varepsilon, \varepsilon + w_2, ..., \varepsilon + w_J] \, d\varepsilon$$

Then:

$$E((\varepsilon_2 - \varepsilon_1)^2 | \delta_1(\varepsilon) = 1)$$

$$= \int_{z_2=-\infty}^{V_1-V_2} z_2^2 \cdot H_2[z_2, V_1 - V_3, ..., V_1 - V_J] \, dz_2$$

Recall that $G[y_1, y_2, ..., y_J] = \alpha y_1 + \alpha y_2 + \alpha G^A[<y_j^A>]$. Thus $G_1 = \alpha$, and by Theorem B.3.1:

$$H[w_2, ..., w_J] = \alpha \left[\alpha + \alpha e^{-w_2/\theta} + \alpha G^A[<e^{-w_j/\theta}>] \right]^{-1}$$

from which follows:

$$H_2[w_2, ..., w_J] = e^{-w_2/\theta} \left[1 + G^A[<e^{-w_j/\theta}>] + e^{-w_2/\theta} \right]^{-2}$$

and:

$$E((\varepsilon_2 - \varepsilon_1)^2 | \delta_1 = 1)$$

$$= \frac{\theta^2}{P_1} \int_{-\infty}^{(V_1-V_2)/\theta} \frac{y^2 e^{-y}}{(A + e^{-y})^2} \, dy$$

$$= \frac{\theta^2}{P_1 A^2} \cdot \int_{-\infty}^{(V_1-V_2)/\theta} \frac{y^2 e^{-y}}{(1 + e^{-\ln A - y})^2} \, dy$$

where $A = 1 + G^A[<e^{-(V_1-V_j)/\theta}>]$ and we have used the transformation $y = z_2/\theta$. Note that:

$$\frac{(1-P_2)}{P_1} = \frac{G - \alpha \, e^{V_2/\theta}}{\alpha \, e^{V_1/\theta}} = \frac{\alpha \, e^{V_1/\theta} + \alpha \, G^A}{\alpha \, e^{V_1/\theta}}$$

$$= 1 + e^{-V_1/\theta} \cdot G^A[<e^{V_j/\theta}>] = A$$

Define $z = y + \ln A$. Then:

$$E(Y_2^2|\delta_1 = 1) = \frac{\theta^2}{P_1 A^2} \cdot \int_{-\infty}^{((V_1-V_2)/\theta) + \ln A} \frac{(z - \ln A)^2 A \, e^{-z}}{(1+e^{-z})^2} \, dz$$

$$= \frac{\theta^2}{P_1 A} \cdot \int_{-\infty}^{((V_1-V_2)/\theta) + \ln A} \frac{(z^2 - 2 z \ln A + (\ln A)^2) \, e^{-z}}{(1+e^{-z})^2} \, dz$$

Because:

$$(V_1-V_2)/\theta = \ln (P_1/P_2) \quad \text{and} \quad A = (1-P_2)/P_1$$

It follows that:

$$(V_1-V_2)/\theta + \ln A = \ln (P_1/P_2) + \ln [(1-P_2)/P_1] = \ln [(1-P_2)/P_2]$$

We let $x = \ln [(1-P_2)/P_2]$, so that:

$$E(Y_{\tilde{2}}^2|\delta_1 = 1) = \frac{\theta^2}{P_1\,A} \int_{-\infty}^{x} \frac{z^2\,e^{-z}}{(1 + e^{-z})^2}\,dz +$$

$$\frac{\theta^2}{P_1\,A} \int_{-\infty}^{x} \frac{e^{-z}}{(1 + e^{-z})^2}\,dz \cdot (\ln A)^2 -$$

$$\frac{2\,(\ln A)\,\theta^2}{P_1\,A} \int_{-\infty}^{x} \frac{e^{z}}{(1 + e^{-z})^2}\,dz$$

$$= \frac{\theta^2}{(1-P_2)} \int_{-\infty}^{x} \frac{z^2\,e^{-z}}{(1 + e^{-z})^2}\,dz + \theta^2\,(\ln\,((1-P_2)/P_1))^2$$

$$\frac{-2\,(\ln A)\,\theta^2}{(1-P_2)} \int_{-\infty}^{x} \frac{z\,e^{-z}}{(1 + e^{-z})^2}\,dz$$

Using integration by parts, we have:

$$\int_{t=-\infty}^{x} \frac{t\,e^{-t}\,dt}{(1 + e^{-t})^2} = \frac{x}{1 + e^{-x}} - \ln\,(1 + e^{x})$$

from which follows:

$$\int_{t=-\infty}^{\ln\,[(1-P_2)/P_2]} \frac{t\,e^{-t}\,dt}{(1 + e^{-t})^2} = \frac{\ln\,[(1-P_2)/P_2]}{(1-P_2)^{-1}} + \ln P_2$$

$$= \left[P_2 \ln P_2 + (1-P_2)\,\ln(1-P_2) \right]$$

Hence:

$$E(Y_{\tilde{2}}^2|\delta_1(\varepsilon) = 1) = \theta^2\,(\ln\,((1-P_2)/P_1))^2$$

$$- \frac{2\,\theta^2}{(1-P_2)} \ln \left((1-P_2)/P_1\right) \left[P_2 \ln P_2 + (1-P_2) \ln (1-P_2) \right]$$

$$+ \frac{\theta^2}{(1-P_2)} \cdot \int_{-\infty}^{\ln \left[(1-P_2)/P_1\right]} h(z)\, dz$$

$$= \theta^2 \left[\ln \left((1-P_2)/P_1\right)\right]^2$$

$$- 2\,\theta^2 \ln \left((1-P_2)/P_1\right) \left[P_2 \ln P_2/(1-P_2) + \ln (1-P_2) \right]$$

$$+ \frac{\theta^2}{(1-P_2)} \cdot \int_{-\infty}^{\ln \left[(1-P_2)/P_2\right]} h(z)\, dz \quad \text{Q.E.D.}$$

The integral $\int_{-\infty}^{x} h(z)\, dz$ where $h(z) = (z^2\, e^{-z})/(1 + e^{-z})^2$ is in fact related to $E[y^2 | y < x]$ where y has a univariate logistic distribution. A closed-form solution for this conditional expectation does not exist. Hay (1980) and Lee (1981) determine the expectation in terms of a series expansion involving the incomplete gamma distribution. Using an alternative series expansion, we derive a computationally simple form for the integral.

Theorem B.3.3. For $0 < \lambda < 1$:

$$\int_0^{\ln \lambda^{-1}} \frac{u^2 \, e^{-u}}{(1 + e^{-u})^2} \, du = \frac{\pi^2}{6} - \frac{\lambda \, (\ln \lambda)^2}{(1 + \lambda)} - 2 \, (\ln \lambda) \, (\ln (1 + \lambda))$$

$$+ 2 \sum_{i=0}^{\infty} (-1)^i \, \frac{\lambda^{i+1}}{(i+1)^2}$$

Proof. From the formula for the sum of a geometric series we have:

$$(1 + x)^{-1} = \sum_{i=0}^{\infty} (-1)^i \, x^i \quad \text{for} \quad |x| < 1$$

Differentiating and integrating, term by term, provides two useful relations:

$$\frac{1}{(1+x)^2} = \sum_{i=1}^{\infty} (-1)^{i+1} \, i \, x^{i-1} = \sum_{i=0}^{\infty} (-1)^i \, (i+1) \, x^i$$

$$\ln (1+x) = \sum_{i=0}^{\infty} \frac{(-1)^i}{(i+1)} \, x^{i+1} \quad \text{for} \quad |x| < 1$$

For $x = e^{-u}$ with $u > 0$:

$$(1 + e^{-u})^{-2} = \sum_{i=0}^{\infty} (-1)^i \, (i+1) e^{-ui} \quad \text{and}$$

$$\int_0^{\ln \lambda^{-1}} \frac{u^2 \, e^{-u}}{(1 + e^{-u})^2} \, du = \int_0^{\ln \lambda^{-1}} u^2 \sum_{i=0}^{\infty} (-1)^i \, (i+1) \, e^{-u(i+1)} \, du$$

$$= \sum_{i=0}^{\infty} (-1)^i \, (i+1) \int_0^{\ln \lambda^{-1}} u^2 \, e^{-u(i+1)} \, du$$

Next we use the fact that:

$$\int y^2 e^{-iy}\, dy = \frac{-1}{i}\left[y^2 + \frac{2}{i}\, y + \frac{2}{i^2} \right] e^{-iy}$$

Then:

$$\int_0^{\ln \lambda^{-1}} \frac{u^2 e^{-u}}{(1+e^{-u})^2}\, du$$

$$= \sum_{i=0}^{\infty} (-1)^i (i+1) \left[\frac{-1}{(i+1)}\, y^2 + \frac{2}{(i+1)}\, y + \frac{2}{(i+1)^2} \right] e^{-(i+1)y} \ \Big|_0^{-\ln \lambda}$$

$$= \sum_{i=0}^{\infty} (-1)^{i+1} \left[\left((\ln \lambda^{-1})^2 + \frac{2}{(i+1)}\ln \lambda^{-1} + \frac{2}{(i+1)^2} \right) \lambda^{i+1} - \frac{2}{(i+1)^2} \right]$$

$$= -2 \sum_{i=0}^{\infty} \frac{(-1)^{i+1}}{(i+1)^2} + (\ln \lambda^{-1})^2 \cdot \sum_{i=0}^{\infty} (-1)^{i+1} \lambda^{i+1}$$

$$+ 2 (\ln \lambda^{-1}) \cdot \sum_{i=0}^{\infty} \frac{(-1)^{i+1}}{(i+1)} \lambda^{i+1} + 2 \sum_{i=0}^{\infty} (-1)^{i+1} \lambda^{i+1}/(i+1)^2$$

$$= \frac{\pi^2}{6} - \left[\frac{\lambda (\ln \lambda)^2}{(1+\lambda)} - 2 (\ln \lambda)(\ln (1+\lambda)) + 2 \sum_{i=0}^{\infty} (-1)^i \lambda^{i+1}/(i+1)^2 \right]$$

where we have used the fact that:

$$\sum_{i=0}^{\infty} (-1)^i/(i+1)^2 = \pi^2/12 \quad \text{Q.E.D.}$$

For reference below, we let:

$$G(\lambda) = \left[\frac{\lambda (\ln \lambda)^2}{(1+\lambda)} - 2 (\ln \lambda) \ln (1+\lambda) + 2 \sum_{i=0}^{\infty} (-1)^i \lambda^{i+1}/(i+1)^2 \right]$$

Application of Theorem B.3.2 in the case of binary alternatives yields:

Corollary B.3.2.

$$E(Y_2^2|\delta_1 = 1) = \begin{cases} (\theta^2/P_1) \cdot (\pi^2/3 - G(P_2/P_1)) & \text{for } P_1 > P_2 \\ (\theta^2/P_1) \cdot (G(P_1/P_2)) & \text{for } P_1 < P_2 \\ (\theta^2/P_1) \cdot (\pi^2/6) & \text{for } P_1 = P_2 \end{cases}$$

Proof. Using Theorem B.3.2:

$$E(Y_2^2|\delta_1 = 1) = \frac{\theta^2}{P_1} \int_{-\infty}^{\ln (P_1/P_2)} h(z) \, dz$$

where we have imposed the restriction $P_1 + P_2 = 1$ implied in the case of binary alternatives. For $P_1 > P_2$:

$$E(Y_2^2|\delta_1 = 1) = \frac{\theta^2}{P_1} \int_{-\infty}^{0} h(z) \, dz + \frac{\theta^2}{P_1} \int_{0}^{\ln (P_1/P_2)} h(z) \, dz$$

Now make the substitution $\lambda^{-1} = P_1/P_2$. Then Theorem B.3.3 implies:

$$E(Y_2^2|\delta_1 = 1) = (\theta^2/P_1)(\pi^2/6) + (\theta^2/P_1) \left[(\pi^2/6) - G(P_2/P_1) \right]$$

For $P_1 < P_2$:

$$E(Y_2^2|\delta_1 = 1) = \frac{\theta^2}{P_1} \int_{-\infty}^{0} h(z) \, dz - \frac{\theta^2}{P_1} \int_{\ln (P_1/P_2)}^{0} h(z) \, dz$$

$$= \frac{\theta^2}{P_1} \frac{\pi^2}{6} - \frac{\theta^2}{P_1} \left[\pi^2/6 - G(P_1/P_2) \right] = \frac{\theta^2}{P_1} G(P_1/P_2)$$

Finally, when $P_1 = P_2$, $G(1) = \pi^2/6$, which implies continuity for $E(Y_2^2$
$|\delta = 1)$ Q.E.D.

B.4. Continuous/discrete econometric systems

We now introduce a random variable η and suppose that conditioned on
ε, η has mean $(\sqrt{6}\sigma/ \pi\theta) \sum_{i-1}^{m} R_i\varepsilon_i$ and variance $\sigma^2 (1 - \sum_{i-1}^{m} R_i^2)$ with
$\sum_{i-1}^{m} R_i = 0$ and $\sum_{i-1}^{m} R_i^2 < 1$.

For the present, we assume that $<\varepsilon_i>$ are independently, identically
extreme value distributed with $E(\varepsilon_i) = 0$. This is accomplished by
assuming that location parameter $\alpha = -\gamma\theta$. Note that $(\sqrt{6}\sigma/\pi\theta) = (\sigma/\sigma_\varepsilon)$
where σ_ε is the standard deviation of ε_i. The next theorem presents the
unconditional moments of η.

Theorem B.4.1. [Dubin and McFadden].

(a) $E(\eta) = 0$

(b) $E(\eta)^2 = \sigma^2$

(c) *Correl* $(\eta, \varepsilon_i) = R_i$

Proof.

(a) $E(\eta) = E[E(\eta|\varepsilon)] = E \left[\frac{\sigma}{\sigma_\varepsilon} \sum_{i-1}^{m} R_i \varepsilon_i \right] = 0$

(b) $E(\eta^2|\varepsilon) = var (\eta|\varepsilon) + (E(\eta|\varepsilon))^2$

$$E(\eta^2) = E \left[\sigma^2 (1 - \sum_{i-1}^{m} R_i^2) + \left[\frac{\sigma}{\sigma_\varepsilon} \sum_{i-1}^{m} R_i \varepsilon_i \right]^2 \right]$$

$$= \sigma^2 \left(1 - \sum_{i=1}^m R_i^2\right) + \frac{\sigma^2}{\sigma_\varepsilon^2} \sum_{i=1}^m R_i^2 \sigma_\varepsilon^2 = \sigma^2$$

(c) $E(\eta\ \varepsilon_i) = E\left[E(\eta\varepsilon_i | \varepsilon)\right] = E\left[\varepsilon_i\ E(\eta | \varepsilon)\right]$

$$= E\left[\varepsilon_i\ \frac{\sigma}{\sigma_\varepsilon} \sum_{i=1}^m R_i\ \varepsilon_i\right] = \frac{\sigma}{\sigma_\varepsilon} R_i\ \sigma_\varepsilon^2 = \sigma\ R_i\ \sigma_\varepsilon$$

Therefore, *Correl* $(\eta,\ \varepsilon_i) = E(\eta\ \varepsilon_i)/\sigma\ \sigma_\varepsilon = R_i$ Q.E.D.

We now derive the expectation of η conditioned on the event that a particular alternative is chosen.

Theorem B.4.2. [Dubin and McFadden].

$$E(\eta | \delta_i(\varepsilon) = 1) = \frac{\sqrt{6}\sigma}{\pi}\left[\sum_{j=1}^m \frac{R_j\ P_j}{(1-P_j)}\ \ln P_j - R_i\ \frac{\ln P_i}{(1-P_i)}\right]$$

Proof. Define $A_i \equiv \{\varepsilon \mid \delta_i(\varepsilon) = 1\}$. Then:

$$E(\eta | \delta_i = 1) = \frac{1}{P_i} \int_{A_i} E(\eta | \varepsilon) \prod_{j=1}^m f(\varepsilon_j)\ d\varepsilon$$

$$= \frac{1}{P_i} \int_{A_i} \left(\frac{\sigma}{\sigma_\varepsilon} \sum_{j=1}^m R_j\ \varepsilon_j\right) \prod_{j=1}^m f(\varepsilon_j)\ d\varepsilon$$

$$= \frac{\sigma}{\sigma_\varepsilon} \sum_{j=1}^m \frac{R_j}{P_i} \int_{A_i} \varepsilon_j \prod_{j=1}^m f(\varepsilon_j)\ d\varepsilon$$

$$= \frac{\sigma}{\sigma_\varepsilon} \sum_{j=1}^m E[\varepsilon_j | \delta_i(\varepsilon) = 1] \cdot R_j$$

$$= \frac{\sigma}{\sigma_\varepsilon} \sum_{j \mid j \neq i}^{m} E[\varepsilon_j \mid \delta_i(\varepsilon) = 1] R_j + \frac{\sigma}{\sigma_\varepsilon} E[\varepsilon_i \mid \delta_i(\varepsilon) = 1] R_i$$

Using Example B.2.1, we find:

$$E(\eta \mid \delta_i(\varepsilon) = 1) = \frac{\sigma}{\sigma_\varepsilon} \sum_{j \mid j \neq i}^{m} \frac{\theta R_j P_j \ln P_j}{(1-P_j)} - \frac{\sigma}{\sigma_\varepsilon} R_i \theta \ln P_i$$

where we have imposed $\alpha = -\gamma\theta$. Noting that $\sigma_\varepsilon = (\pi \theta/\sqrt{6})$, we have:

$$E(\eta \mid \delta_i(\varepsilon) = 1) = \frac{\sqrt{6}\sigma}{\pi} \left[\left(\sum_{j \mid j \neq i}^{m} \frac{R_j P_j \ln P_j}{(1-P_j)} \right) - R_i \ln P_i \right]$$

$$= \frac{\sqrt{6}\sigma}{\pi} \left[\left(\sum_{j-1}^{m} \frac{R_j P_j \ln P_j}{(1-P_j)} \right) - \frac{R_i \ln P_i}{(1-P_i)} \right] \quad \text{Q.E.D.}$$

Let $\delta_{ij} = 1$ if $i = j$ and 0 otherwise. Then we may rewrite the result of Theorem B.4.2 as:

$$E(\eta \mid \delta_i(\varepsilon) = 1) = \frac{\sqrt{6}\sigma}{\pi} \left[\left(\sum_{j \mid j \neq i}^{m} \frac{R_j P_j \ln P_j}{(1-P_j)} \right) + \frac{R_i \ln P_i (P_i-1)}{(1-P_i)} \right]$$

$$= \frac{\sqrt{6}\sigma}{\pi} \left[\sum_{j-1}^{m} \frac{R_j \ln P_j}{(1-P_j)} (P_j - \delta_{ij}) \right]$$

We now consider the conditional variance of η for the binary case $m = 2$. Recall that:

$$E(\eta^2 \mid \delta_i = 1) = \frac{1}{P_i} \int_{A_i} E(\eta^2 \mid (\varepsilon)) f(\varepsilon) \, d\varepsilon$$

where $f(\varepsilon) = \prod_{i-1}^{m} f(\varepsilon_i)$. We use the relation:

$$E(\eta^2|\varepsilon) = var(\eta|\varepsilon) + (E(\eta|\varepsilon))^2$$

$$= \sigma^2 \left(1 - \sum_{i=1}^{2} R_i^2\right) + \frac{\sigma^2}{\sigma_\varepsilon^2} \left(\sum_{i=1}^{2} R_i \, \varepsilon_i\right)^2$$

to obtain:

$$E(\eta^2|\delta_i = 1) = \sigma^2 \left(1 - \sum_{i=1}^{2} R_i^2\right) + \frac{\sigma^2}{\sigma_\varepsilon^2} \sum_{i=1}^{2} R_i^2 \, E(\varepsilon_i^2|\delta_i = 1)$$

$$+ \frac{2 \, \sigma^2}{\sigma_\varepsilon^2} R_1 \, R_2 \, E(\varepsilon_1 \, \varepsilon_2|\delta_i = 1)$$

We collect results in the next theorem:

Theorem B.4.3. [Dubin and McFadden].

$$E(\eta^2|\delta_1) = \sigma^2 + 2 \, \sigma^2 \, R_2^2 \, J(P_1,\delta_1) \quad \text{where} \quad J(P_1, \delta_1) =$$

$$\left\{
\begin{array}{ll}
1/P_1 - 1 - (3/\pi^2) \, (1/P_1) \cdot G\left[\dfrac{(1-P_1)}{P_1}\right] & \text{for } \delta_1 = 1, P_1 > 1/2 \\[2ex]
- 1 + (3/\pi^2) \, (1/P_1) \cdot G\left[\dfrac{P_1}{(1-P_1)}\right] & \text{for } \delta_1 = 1, P_1 \leqslant 1/2 \\[2ex]
- 1 + (3/\pi^2) \, [1/(1-P_1)] \cdot G\left[\dfrac{(1-P_1)}{P_1}\right] & \text{for } \delta_1 = 0, P_1 > 1/2 \\[2ex]
P_1/(1-P_1) - (3/\pi^2) \, [1/(1-P_1)] \cdot G\left[\dfrac{P_1}{(1-P_1)}\right] & \text{for } \delta_1 = 0, P_1 \leqslant 1/2
\end{array}
\right.$$

Proof.

$$E[\eta^2|\delta_1 = 1] = \sigma^2 (1 - (R_1^2 + R_2^2)) + \frac{\sigma^2}{\sigma_\varepsilon^2} \left[R_1^2 \ E(\varepsilon_1^2|\delta_1 = 1) \right.$$

$$\left. + R_2^2 \ E(\varepsilon_2^2|\delta_1 = 1) + 2 \ R_1 \ R_2 \ E(\varepsilon_1\varepsilon_2|\delta_1 = 1) \right]$$

For the binary case: $P_1 + P_2 = 1$ and $R_1 + R_2 = 0$. Example B.2.1 and Corollary B.3.2 then imply:

$$E(\eta^2|\delta_1 = 1) = \sigma^2 (1 - 2 \ R_2^2)$$

$$+ (\sigma_2^2/\sigma_\varepsilon^2) \ R_2^2 \left\{ \begin{array}{ll} \dfrac{\theta^2}{P_1} \left[(\pi^2/3) - G\left(\dfrac{(1-P_1)}{P_1}\right)\right] & \text{for } P_1 > 1/2 \\[20pt] \dfrac{\theta^2}{P_1} \ G\left(\dfrac{P_1}{(1-P_1)}\right) & \text{for } P_1 \leqslant 1/2 \end{array} \right.$$

Using $\sigma_\varepsilon^2 = \pi^2 \ \theta^2/6$ yields the first two parts of the theorem. It is then simple to derive the expression for $E(\eta^2|\delta_1 = 0)$ from:

$$E(\eta^2|\delta_1 = 1) \ P_1 + E(\eta^2|\delta_1 = 0) \ (1-P_1) = E(\eta^2) = \sigma^2 \quad \text{Q.E.D.}$$

If we relax the assumption that $<\varepsilon_i>$ are independently, identically extreme value distributed, then conditional moments of η are not easily derived. Indeed, the strong separability used for the function G in Corollary B.2.2 and Theorem B.3.2 if applied symmetrically to all components of G would imply the simple multinomial logit specification.

The sequential form of the GEV family does, however, provide a tractable alternative. Rather than assume that η has a linear conditional expectation in $<\varepsilon_i>$, we instead assume that η has a linear conditional expectation in the space of the "induced" independent extreme value random variables that generate the conditional branch probabilities. This assumption is motivated by the consideration that the simple multinomial logit probability form is implied by but does not itself imply an independent extreme value error structure. This point is usefully illustrated in the bivariate extreme value distribution:

$$G(y) = \left[y_1^{1/(1-\sigma)} + y_2^{1/(1-\sigma)} \right]^{(1-\sigma)} \tag{53}$$

Equation (53) implies a probability choice system:

$$P_1 = \frac{e^{V_1/\theta(1-\sigma)}}{e^{V_1/\theta(1-\sigma)} + e^{V_2/\theta(1-\sigma)}} \tag{54}$$

Alternatively, consider the independent form of the GEV:

$$G[y] = y_1 + y_2 \tag{55}$$

which implies the multinomial logit probability choice system:

$$P_1 = \frac{e^{V_1/\theta}}{e^{V_1/\theta} + e^{V_2/\theta}} \tag{56}$$

As the scale parameters $\theta(1-\sigma)$ and θ are *not* identified in (54) and (56), the resulting models are observationally equivalent.

    Furthermore, in the sequential or nested logit model, we may view the second-level conditional probabilities in equation (47) as being generated by the independent extreme value random variables $<\varepsilon_j^{B_m}>$ with variance $(\pi^2/6)(1 - \sigma_m)^2$. Specifically:

$$P[i \mid B_m] = Prob \; [V_i + \varepsilon_i^{B_m} \geq V_j + \varepsilon_j^{B_m} , \; \forall j \mid j \in B_m \text{ and } j \neq i] \tag{57}$$

Finally, the error structure $<\varepsilon_j^{B_m}>$ may be analyzed through Theorems B.4.1, B.4.2, and B.4.3.

References

Achenback, P. and C. Coblentz (1963), "Field Measurements of Air Infiltration in Ten Electrically-Heated Houses," ASHRAE Annual Meeting, Paper No. 1845.

Acton, J. P., B. M. Mitchell and R. S. Mowill (1976), *Residential Demand for Electricity in Los Angeles: An Econometric Study of Disaggregated Data*, The Rand Corporation, R-1899-NSF.

Acton, J. P., B. M. Mitchell and R. S. Mowill (1978), *Residential Electricity Demand under Declining Block Tariffs: An Econometric Study Using Micro-Data*, The Rand Corporation, P-6203.

Amemiya, T. (1973), "Regression Analysis When the Dependent Variable is Truncated Normal," *Econometrica* 41, 997-1016.

Amemiya, T. (1974a), "Bivariate Probit Analysis: Minimum Chi-Square Methods," *Journal of the American Statistical Association* 69, 940-944.

Amemiya, T. (1974b), "Multivariate Regression and Simultaneous Equation Models When the Dependent Variables are Truncated Normal," *Econometrica* 42, 999-1012.

Amemiya, T. (1974c), "A Note on the Fair and Jaffee Model," *Econometrica* 42, 759-762.

Amemiya, T. (1978a), "The Estimation of a Simultaneous Equation Generalized Probit Model," *Econometrica* 46, 1193-1205.

Amemiya, T. (1978b), "On a Two-Step Estimation of a Multivariate Logit Model," *Journal of Econometrics* 8, 13-21.

Amemiya, T. (1979), "The Estimation of a Simultaneous Equation Tobit Model," *International Economic Review* 20, 169-181.

Amemiya, T. and G. Sen (1977), "The Consistency of the Maximum Likelihood Estimator in a Disequilibrium Model," Tech. Report 238, Institute for Math. Studies in Social Sciences, Stanford University, Stanford.

American Society of Heating, Refrigerating, and Air-Conditioning Engineering (1977), *Handbook and Product Directory, Fundamentals*, New York: ASHRAE.

American Society of Heating, Refrigerating, and Air-Conditioning Engineering (1978), *Handbook and Product Directory, Applications*, New York: ASHRAE.

American Society of Heating, Refrigerating, and Air-Conditioning Engineering (1979), *Handbook and Product Directory, Equipment*, New York: ASHRAE.

Anderson, K. P. (1973), *Residential Energy Consumption: Single Family Housing*, U. S. Department of Commerce, Publ. No. HUD-PDR-29.2.

Annual Climatological Data—National Summary 1978-1979.

Balestra, P. and M. Nerlove (1966), "Pooling Cross Section and Time Series Data in the Estimation of a Dynamic Model: The Demand for Natural Gas," *Econometrica* 34, 585-612.

Barnes, R., R. Gillingham and R. Hagemann (1981), "The Short-Run Residential Demand for Electricity," *Review of Economics and Statistics* 63, 541-552.

Becker, G. S. (1965). "A Theory of the Allocation of Time," *Economic Journal* 75, 493-517.

Berg, S. and W. Roth (1976), "Some Remarks on Residential Electricity Consumption and Social Rate Restructuring," *Bell Journal of Economics* 7, 690-698.

Berger, A. (1981), *Building Design and Cost File*, Vol. I (General Construction Trades), Vol. II (Mechanical and Electrical Trades), New York: Van Nostrand.

Berndt, E. R. (1978), "The Demand for Electricity: Comment and Further Results," Resources Paper No. 28, Department of Economics, University of British Columbia.

Billings, R. B. (1982), "Specification of Block Rate Price Variables in Demand Models," *Land Economics* 58, 385-394.

Billings, R. B. and D. E. Agthe (1980), "Price Elasticities for Water: A Case of Increasing Block Rates," *Land Economics* 56, 73-84.

Blackorby, C., G. Lady, D. Nissen and R. R. Russell (1970), "Homothetic Separability and Consumer Budgeting," *Econometrica* 38, 468-472.

Blumstein, C., C. York and W. Kemp (1979), "An Assessment of the National Interim Energy Consumption Survey," Energy Resources Group, University of California, Berkeley.

Boyce, W. and R. DiPrima (1969), *Elementary Differential Equations and Boundary Value Problems*, New York: John Wiley and Sons, Inc.

Brownstone, D. (1980), *An Econometric Model of Consumer Durable Choice and Utilization Rate*, unpublished Ph.D. thesis, University of California, Berkeley.

Burtless, G. and J. Hausman (1978), "The Effect of Taxation on Labor Supply," *Journal of Political Economy* 86, 1103-30.

California State Energy Commission (1979), "California Energy Demand 1978-2000," Working Paper.

Cambridge Systematics/West (1979), "An Analysis of Household Survey Data in Household Time-of-Day and Annual Electricity Consumption," Cambridge Systematics/West, Working Paper.

Cambridge Systematics, Inc. (1981), "Development of a Residential Appliance Survey for New England Electric System," Working Paper.

Chern, W. S. and W. Lin (1976), "Energy Demand for Space Heating: An Econometric Analysis," *American Statistical Association, Proceedings of the Business and Economic Statistics Section*.

Chipman, J. S., L. Hurwicz, M. K. Richter and H. F. Sonnenschein (1971), *Preferences, Utility, and Demand*, New York: Harcourt Brace, Jovanovich.

Chow, G. C. (1957), "Contribution to Economic Analysis," *Demand for Automobiles in the United States*, Vol. XIII, Amsterdam: North-Holland Publishing Company.

Cowing, T. (1974), "Technical Change and Scale Economies in an Engineering Production Function: The Case of Steam Electric Power," *Journal of Industrial Economics* 23, 135-152.

Cowing, T., J. Dubin and D. McFadden (1981), "Residential Energy Demand Modeling and the NIECS Data Base: An Evaluation," Massachusetts Institute of Technology Energy Laboratory Report No. MIT-EL-82-009.

Cowing, T., J. Dubin and D. McFadden (1982), "The NIECS Data Base and Its Use in Residential Energy Demand Modeling," Energy Laboratory Discussion Paper No. 24, MIT-EL 82-041WP Massachusetts Institute of Technology.

Cragg, J. G. (1971), "Some Statistical Models for Limited Dependent Variables with Application to the Demand for Durable Goods," *Econometrica* 39, 829-844.

Cragg, J. G. and R. Uhler (1970), "The Demand for Automobiles," *Canadian Journal of Economics* 3, 386-406.

Diewert, W. D. (1974), "Intertemporal Consumer Theory and the Demand for Durables," *Econometrica* 42, 497-516.

Dubin, J. A. (1982), "Economic Theory and Estimation of the Demand for Consumer Durable Goods and their Utilization: Appliance Choice and the Demand for Electricity," Ph.D. Dissertation, Massachusetts Institute of Technology.

Dubin, J. A. (1983), "The National Interim Energy Consumption Survey (NIECS) and the Pacific Northwest Data Base (PNW)—A Summary and Collected Programs," California Institute of Technology Working Paper.

Dubin, J. A. and D. L. McFadden (1983), "A Heating and Cooling Load Model for Single-Family Detached Dwellings in Energy Survey Data," California Institute of Technology Working Paper No. 469.

Dubin, J. A. and D. L. McFadden (1984), "An Econometric Analysis of Residential Electric Appliance Holdings and Consumption," *Econometrica* 52, 345-362.

Duncan, G. (1980a), "Formulation and Statistical Analysis of the Mixed Continuous Discrete Variable Model in Classical Production Theory," *Econometrica* 48, 839-851.

Duncan, G. (1980b), "Mixed Continuous Discrete Choice Models in the Presence of Hedonic or Exogenous Price Functions," Washington State University, Working Paper.

Duncan, G. and D. Leigh (1980), "Wage Determination in the Union and Non-Union Sectors: A Sample Selectivity Approach," *Industrial and Labor Relations Review* 34, 24-34.

Engelsman, C. (1981), *Residential Cost Manual*, New York: Van Nostrand.

Fair, R. C. and D. M. Jaffee (1972), "Methods of Estimation for Markets in Disequilibrium," *Econometrica* 40, 497-514.

Fisher, F. M. and C. Kaysen (1962), *A Study in Econometrics: The Demand for Electricity in the U.S.*, Amsterdam: North-Holland Publishing Company.

Foster, H. S. and B. R. Beattie (1981), "On the Specification of Price in Studies of Consumer Demand Under Block Price Scheduling," *Land Economics* 57, 624-629.

Fuss, M. and D. McFadden (1978), "Flexibility versus Efficiency in Ex-Ante Plant Design," in: Fuss and McFadden, eds., *Production Economics: A Dual Approach to Theory and Applications*, Amsterdam: North-Holland Publishing Company.

Geary, P. T. and M. Morishima (1973), "Demand and Supply under Separability," in: M. Morishima and others, eds., *Theory of Demand: Real and Monetary*, Oxford: Clarendon.

George, S. (1979), Short-Run Residential Electricity Demand: A Policy Oriented Look, Ph.D. Dissertation, University of California, Davis.

George, S. (1981), "A Review of the Conditional Demand Approach to Electricity Demand Estimation," (revised version of a paper given at the EPRI Conference on "End-Use Modeling and Conservation Analysis," Atlanta, Georgia).

Goett, A. (1979), "A Structured Logit Model of Appliance Investment and Fuel Choice," Working Paper, Cambridge Systematics, Inc./West.

Goett, A., D. McFadden and G. L. Earl (1980), "The Residential End-Use Energy Policy System: Model Description and Analysis of Energy Usage and Efficiency Choice."

Goldfeld, S. M. and R. E. Quandt (1972), *Nonlinear Methods on Econometrics*, Amsterdam: North-Holland Publishing Company.

Goldfeld, S. M. and R. E. Quandt (1973), "The Estimation of Structural Shifts by Switching Regressions," *Annals of Economic and Social Measurement* 2, 475-485.

Goldfeld, S. M. and R. E. Quandt (1976), "Techniques for Estimating Switching Regressions," in: S. M. Goldfeld and R. E. Quandt, eds., *Studies in Non-Linear Estimation*, Cambridge: Ballinger.

Gorman, W. M. (1959), "Separable Utility and Aggregation," *Econometrica* 27, 469-481.

Griffin, A.H. and W. E. Martin (1981), "Price Elasticities for Water: A Case of Increasing Block Rates: Comment," *Land Economics* 57, 266-275.

Griliches, Z. (1960), "The Demand for a Durable Input: Farm Tractors in the United States, 1921-57," in: Harberger ed., *Demand for Durable Goods*, Chicago, Illinois: University of Chicago Press, 181-207.

Hall, R. E. (1973), "Wages, Income, and Hours of Work in the U.S. Labor Force," in: Cain and Watts, eds., *Income Maintenance and Labor Supply: Econometric Studies*, Chicago: Rand McNally.

Halvorsen, R. (1982), *Econometric Models of U.S. Energy Demand*, Lexington, Mass.: D.C. Heath.

Hartley, M. J. (1977), *On the Estimation of a General Switching Regime Model Via Maximum Likelihood Methods*, State University of New York at Buffalo, Department of Economics Discussion Paper No. 415

Hartman, R. S. (1978), *A Critical Review of Single Fuel and Interfuel Substitution Residential Energy Demand Models*, Massachusetts Institute of Technology Energy Laboratory Report, MIT-EL-78-003.

Hartman, R. S. (1979a), "Discrete Consumer Choice Among Alternative Fuels and Technologies for Residential Energy Using Appliances," Massachusetts Institute of Technology Energy Laboratory Working Paper, MIT-EL-79-049WP.

Hartman, R. S. (1979b), "Frontiers in Energy Demand Modeling," *Annual Review of Energy* 4, 433-466.

Hartman, R. S. and A. Werth (1981), "Short-Run Residential Demand for Fuels: A Disaggregated Approach," *Land Economics* 57,197-212.

Hausman, J. (1978), "Specification Tests in Econometrics," *Econometrica* 46, 1251-1271.

Hausman, J. (1979), "Individual Discount Rates and the Purchase and Utilization of Energy-Using Durables," *The Bell Journal of Economics* 10, 33-54.

Hausman, J. (1981a), "Exact Consumer's Surplus and Deadweight Loss," *American Economic Review* 71, 663-676.

Hausman, J. (1981b), "Stochastic Problems in the Simulation of Labor Supply," Massachusetts Institute of Technology, Department of Economics Working Paper.

Hausman, J., M. Kinnucan and D. McFadden (1979), "A Two-Level Electricity Demand Model," *Journal of Econometrics* 10, 263-289.

Hausman, J. and D. McFadden (1984), "Specification Tests for the Multinomial Logit Model," *Econometrica* 52, 1219-1240.

Hausman, J. and W. E. Taylor (1981), "Panel Data and Unobservable Effects," *Econometrica* 49, 1377-1393.

Hausman, J. and J. Trimble (1981), "Appliance Purchase and Usage Adaptation to a Permanent Time of Day Electricity Rate Schedule," mimeo, Massachusetts Institute of Technology.

Hay, J. (1980), "An Analysis of Occupational Choice and Income," Ph.D. Dissertation, Yale University.

Heckman, J. (1976a), "Simultaneous Equations Model with Continuous and Discrete Endogenous Variables and Structural Shift," in: Goldfeld and Quandt, eds., *Studies in Non-Linear Estimation*, Cambridge: Ballinger.

Heckman, J. (1976b), "The Common Structure of Statistical Models of Truncation, Sample Selection and Limited Dependent Variables and a Simple Estimation for Such Models," *Annals of Economic and Social Measurement* 5, 475-492.

Heckman, J. (1978), "Dummy Endogenous Variables in a Simultaneous Equation System," *Econometrica* 46, 931-959.

Heckman, J. (1979), "Sample Selection Bias as a Specification Error," *Econometrica* 47, 153-161.

Hildreth, C. and J. P. Hauck (1968), "Some Estimators for a Linear Model with Random Coefficients," *Journal of the American Statistical Association* 63, 584-95.

Hirst, E. and J. Carney (1978), "The Oak Ridge National Laboratory Engineering-Economic Model of Residential Energy Use," ORNL/CON-24.

Hirst, E., R. Goeltz and J. Carney (1981), "Residential Energy Use and Conservation Actions: Analysis of Disaggregate Household Data," ORNL/CON-68. Oak Ridge National Laboratory.

Houthakker, H. S. (1951a), "Electricity Tariffs in Theory and Practice," *Economic Journal* 61, 1-25.

Houthakker, H.S. (1951b), "Some Calculations of Electricity Consumption in Great Britain," *JRSS Series A* 114, 351-371.

Houthakker, H. S. and L. Taylor (1970), *Consumer Demand in the U.S.*, 2nd edition, Cambridge, Mass.: Harvard University Press.

Hyum, D. (1974), *Preliminary Cost Guide*, Pasadena, Calif.: Architectural Data Corporation.

Joskow, P. L. and F. S. Mishkin (1977), "Electric Utility Fuel Choice Behavior in the United States," *International Economic Review* 18, 719-736.

Khashab, A. (1977), *HVAC Estimating Manual*, New York: McGraw-Hill.

King, M. (1980), "An Econometric Model of Tenure Choice and Demand for Housing as a Joint Decision," University of Birmingham Working Paper.

Lancaster, K. (1966), "A New Approach to Consumer Theory," *Journal of Political Economy* 74, 132-57.

Lee, L. F. (1978), "Unionism and Wage Rates: A Simultaneous Equation Model with Qualitative and Limited Dependent Variables," *International Economic Review* 19, 415-433.

Lee, L. F. (1981), "Simultaneous Equations Models with Discrete and Censored Variables," in: C. Manski and D. McFadden, eds., *Structural Analysis of Discrete Data*, Cambridge, Mass.: The Massachusetts Institute of Technology Press.

Lee, L. F. (1982), "Some Approaches to the Correction of Selectivity Bias", *Review of Economic Studies* 40, 355-371.

Lee, L. F. and R. P. Trost (1978), "Estimation of Some Limited Dependent Variable Models with Applications to Housing Demand," *Journal of Econometrics* 8, 357-382.

Lee, L. F., G. S. Madalla and R. Trost (1980), "Asymptotic Convariance Matrices of Two-Stage Probit and Two-Stage Tobit Methods for Simultaneous Equations Models with Selectivity," *Econometrica* 48, 491-503.

Li, M. (1977), "A Logit Model of Home Ownership," *Econometrica* 45, 1081-1097.

Maddala, G. and F. Nelson (1974), "Maximum Likelihood Methods for Markets in Disequilibrium," *Econometrica* 42, 1013-1030.

Maddala, G. and F. Nelson (1975), "Switching Regression Models with Exogenous and Endogenous Switching," *Proceedings of the American Statistical Association*, Business and Economics Section, 423-426.

Mardia, K. V. (1970), *Families of Bivariate Distributions*, London: Griffin.

McFadden, D. (1973), "Conditional Logit Analysis of Qualitative Choice Behavior," *Frontiers in Econometrics*, New York: Academic Press.

McFadden, D. (1974), "The Measurement of Urban Travel Demand," *Journal of Public Economics* 3, 303-328.

McFadden, D. (1978), "Modeling the Choice of Residential Location," *Spatial Interaction Theory and Residential Location*, Amsterdam: North-Holland Publishing Company, 75-96.

McFadden, D. (1979a), "Econometric Net Supply Systems for Firms with Continuous and Discrete Commodities," Massachusetts Institute of Technology, Department of Economics Working Paper.

McFadden, D. (1979b), "Quantitative Methods for Analyzing Travel Behavior of Individuals," in D. Hensher and P. Stopher, eds., *Behavioral Travel Modeling*, Croom-Helm.

McFadden, D. (1981a), "An Evaluation of the ORNL Residential Energy Use Model," Draft Report, Massachusetts Institute of Technology Energy Laboratory, Energy Model Analysis Program.

McFadden, D. (1981b), "Econometric Models of Probabilistic Choice," in: C. Manski and D. McFadden, eds., *Structural Analysis of Discrete Data*, Cambridge, Mass.: Massachusetts Institute of Technology Press.

McFadden, D., D. Kirshner and C. Puig (1978), "Determinants of the Long-Run Demand for Electricity," *Proceedings of the American Statistical Association*.

McFadden, D., W. Tye and K. Train (1978), "Diagnostic Tests for the Independence from Irrelevent Alternatives Property of the Multinomial Logit Model," Transportation Research Record.

McFadden, D. and C. Winston (1981), "Joint Estimation of Discrete and Continuous Choices in Freight Transportation," Working Paper, Department of Economics, Massachusetts Institute of Technology.

Means, R. (1981), *Building Construction Cost Data*, Kingston, Mass.: Means.

Means, R. (1981), *Mechanical and Electrical Cost Data*, Kingston, Mass.: Means.

Moselle, G. (1981), *Building Cost Manual*, Solana Beach, Calif.: Craftsman.

Mount, T. D., L. D. Chapman and T. J. Tyrell (1973), "Electricity Demand in the United States: An Econometric Analysis," Oak Ridge National Laboratory.

Muellbauer, J. (1974), "Household Production Theory, Quality, and the 'Hedonic Technique'," *American Economic Review* 64, 977-994.

Murray, M. P. (1978), "The Demand for Electricity in Virginia," *Review of Economics and Statistics* 60, 585-638.

Muth, R. F. (1966), "Household Production and Consumer Demand Functions," *Econometrica* 34, 699-708.

NAHB (1978), *Thermal Performance Guidelines*, Washington: National Association of Home Builders.

National Electric Rate Book, August 1978, DOE/EIA-0110 (78), Department of Energy, Energy Data Reports, New Jersey.

Nordin, J. A. (1976), "A Proposed Modification of Taylor's Demand Analysis: Comment," *Bell Journal of Economics* 7, 719-721.

Ohta, M. (1975), "Production Technologies of the U.S. Boiler and Turbo-generator Industries and Hedonic Price Indexes for their Products: A Cost Function Approach," *Journal of Political Economy* 83, 1-25.

Olson, R. J. (1980), "A Least Squares Correction for Selectivity Bias," *Econometrica* 48, 1815-1820.

Parti, M. and C. Parti (1980), "The Total and Appliance-Specific Conditional Demand for Electricity in the Household Sector," *Bell Journal of Economics* 11, 309-321.

Pereira, P. ed. (1981), *Dodge Building Cost Calculator and Valuation Guide*, New York: McGraw-Hill.

Pesaran, M. H. (1975), "On the General Problem of Model Selection," *Review of Economic Studies* 41, 153-171.

Pollak, R. A. and M. L. Wachter (1975), "The Relevance of the Household Production Function and Its Implications for the Allocation of Time," *Journal of Political Economy* 83, 255-277.

Pulver, H. (1969), *Construction Estimates and Costs*, New York: McGraw-Hill.

Quandt, R. E. (1972), "A New Approach to Estimating Switching Regressions," *Journal of the American Statistical Association* 67, 306-310.

Response Analysis Corporation (1976), *Lifestyles and Household Energy Use: Report on Methodology*, No. 3819, Princeton, New Jersey.

Response Analysis Corporation (1981), *Report on Methodology*, 6 vols., Princeton, New Jersey.

Rosen, S. (1974), "Hedonic Prices and Implicit Markets: Product Differentiations in Pure Competition," *Journal of Political Economy* 82, 34-55.

Samuelson, P. A. (1950), "The Problem of Integrability in Utility Theory," *Econometrica* 17, 355-85.

Sarviel, E. editor (1981), *National Construction Estimator*, Solana Beach, Calif.: Craftsman.

Shoemaker, M. (1980), *Building Estimator's Reference Book*, Chicago: Walker.

Streeter, R. (1966), "Guide for Estimating Ultimate Cost of Heating Systems," Institute for Building Research, Pennsylvania State University.

Taylor, L. D. (1975), "The Demand for Electricity: A Survey," *Bell Journal of Economics* 6, 74-110.

Taylor, L. D., G. R. Blattenberger and P. K. Verleger (1977), "The Residential Demand for Energy," Palo Alto, Calif.: Electric Power Research Institute.

Terza, J. V. and W. P. Welch (1982), "Estimating Demand Under Block Rates: Electricity and Water," *Land Economics* 58, 181-188.

U.S. Department of Energy, Energy Information Administration, Office of Energy Markets and End Use, DOE/EIA-0272, *Exploring the Variability in Energy Consumption*, July 1981.

U.S. Department of Energy, Energy Information Administration, Office of Energy Markets and End Use, DOE/EIA-0272/5, *Exploring the Variability in Energy Consumption: A Supplement*, October 1981.

U.S. Department of Energy, Energy Information Administration, *Residential Energy Consumption Survey: Characteristics of the Housing Stocks and Households, 1978*, February 1980, DOE/EIA-0207/2.

U.S. Department of Energy, Energy Information Administration, *Residential Energy Consumption Survey: Conservation*, February 1980, DOE/EIA-0207/3.

U.S. Department of Energy, Energy Information Administration, *Residential Energy Consumption Survey: Consumption and Expenditures, April 1979 Through March 1979*, July 1980. DOE/EIA- 1017/5.

U.S. Department of Energy, Energy Information Administration, *Residential Energy Consumption Survey: 1979-1980 Consumption and Expenditures, Part I: National Data (Including Conservation)*, April 1981, DOE/EIA-0262/1.

U.S. Department of Energy, Energy Information Administration, *Residential Energy Consumption Survey: 1979-1980 Consumption and Expenditures, Part II: Regional Data*, May 1981, DOE/EIA-0262/2.

U.S. Department of Energy, Energy Information Administration, *Single-Family Households: Fuel Inventories and Expenditures: National Interim Energy Consumption Survey*, December 1979, DOE/EIA-0207/1.

U.S. Department of Energy, "State Energy Fuel Prices by Major Economic Sectors from 1960-1977," DOE/EIA-0190, July 1979.

U.S. Department of Energy, Office of Energy Markets and End Use Energy Information Administration, "Technical Documentation for the Residential Energy Consumption Survey: National Interim Energy Consumption Survey 1978-1979, Household Monthly Energy Consumption and Expenditures—Public Use Data Tapes—User's Guide," August, 1981.

Werth, A. (1978), "Residential Demand for Electricity and Natural Gas in the Short Run: An Econometric Analysis," Massachusetts Institute of Technology Energy Laboratory Working Paper No. MIT-EL-78-031WP.

Wilder, R. P. and J. F. Willenborg (1975), "Residential Demand for Electricity: A Consumer Panel Approach," *Southern Economic Journal* 42, 212-217.

Wills, H. R. (1977), "Estimating Input Demand Equations by Direct and Indirect Methods," *Review of Economic Studies* 48, 255-270.

Wilson, J. (1971), "Residential Demand for Electricity," *Quarterly Review of Economics and Business* 11, 7-19.

Wu, D. (1973), "Alternative Tests of Independence Between Stochastic Regressors and Disturbances," *Econometrica* 41, 733-750.

Index